Guide to Networking for Physical Security Systems

Guide to Networking for Physical Security Systems

by David J. Engebretson

SECURITY NETWORKING INSTITUTE

THOMSON

DELMAR LEARNING

Australia • Canada • Mexico • Singapore • Spain • United Kingdom • United States

THOMSON

DELMAR LEARNING

Guide to Networking for Physical Security Systems

David J. Engebretson

Vice President, Technology Professional Business Unit:
Gregory L. Clayton

Director of Learning Solutions:
Sandy Clark

Managing Editor:
Larry Main

Acquisitions Editor:
Ed Francis

Marketing Director:
Beth A. Lutz

Channel Manager:
Taryn Zlatin

Marketing Specialist:
Marissa Maiella

Marketing Coordinator:
Jennifer Stall

Product Manager:
Stephanie Kelly

Editorial Assistant:
Jaclyn Ippolito

Director of Production:
Patty Stephan

Production Manager:
Andrew Crouth

Content Project Manager:
Benj Gleeksman

Cataloging-in-Publication Data is on file with the Library of Congress.

ISBN: 1-4180-7396-2

NOTICE TO THE READER

Dedication

This book is dedicated to my mother, Anne, who encouraged me to write, and to the memory of my father, John, who was a living monument to manly virtue.

Contents

Acknowledgments xix
About the Author xx
Introduction xxi

Chapter 1   The Beginnings of Networking 1
             The history in brief 1
             How did these networks develop? 1
             Telephone networks 2
             Mainframe computer networks 3
             Cabling system standardization 4
             Personal computers 4
             Network protocols 5
             The rise of the Internet 6
             Network history and electronic security 6

Chapter 2   Analog and Digital Communications 7
             Communication methods 7
             Analog communications 7
             Problems with analog communications 7
             Digital communications—zeros and ones 9
             Interference 10
             Attenuation 10
             Signal power 10
             Summary 10

Chapter 3   Overview of Ethernet LANs 11
             Evolution 11
             Overview 11
             Cabling 12
             Computer hardware 12
             Firmware 12
             Individual addressing 12
             Nodes 13
             Segments 13
             How transmission happens 13
             What about collisions? 13
             Ethernet data throughput standards 13

Ethernet goes turbo 14
Ethernet and electronic security 14

Chapter 4   Basics of Ethernet 15
Why data is reformatted for transmission 15
Ethernet packet format 15
Packet routing 16
Device addressing 16
MAC addressing 17
IP addressing 17
Network classes and the Internet 18
Class A networks 18
Class B networks 18
Class C networks 18
Reuse of IP addresses 19
Default IP addresses 19
Addresses and communication 19
Why two types of addresses—IP and MAC? 20
Subnet masking 20
Summary 20

Chapter 5   Ethernet Devices and Components 21
Ethernet devices 21
Proxy server 26
Summary 26

Chapter 6   Wire and Cable 27
Initial cabling methods 27
Rethinking the concept 27
Star configuration 28
568 standards overview 29
    Type of cable 29
    Types of connectors 29
    Maximum distances of cable runs 32
    Testing methods 33
Ethernet copper connections 33
UTP patch cords and jumpers 34
Crossover cables 34
Copper cable performance enhancements—all those "Cats" 34
Copper cable installation and performance 35
Summary 35

Chapter 7   Fiber Optics 37
Why fiber is used in networking 37
How fiber works 37

Types of fiber and their uses 39
Fiber connections 39
Fiber testing 39
Electrical-to-fiber converters 40
Analog transmission on fiber 40
Ethernet media converters 41
Selection of media converters 42
    When to use media converters for security applications 43
Powering devices 43
Summary 43

Chapter 8   Wireless LANs 45
The history lesson 45
The concepts of Wi-Fi 45
    Use of Shared Frequencies 46
    Emulation of Ethernet 46
    Standardization 46
Product compatibility 47
Data security 47
Wi-Fi components 47
Access points 47
Wi-Fi laptops 48
Wi-Fi routers and switches 48
Ad hoc mode 48
Wi-Fi coverage 49
Wi-Fi security concerns 50
Bandwidth realities 51
Typical Wi-Fi uses for electronic security 52
The future of Wi-Fi electronic security 53
Summary 53

Chapter 9   IP Addressing Technologies 55
Host 55
Server 55
Static IP 55
Dynamic host configuration protocol—DHCP 56
Static vs. DHCP in electronic security applications 57
Domain name server—DNS 58
Dynamic domain name server—DDNS 58
Ports & network address translation—NAT 59
Summary 61

Chapter 10   Internet WAN Connections and Services. 63
What is the Internet? 63
Internet service providers 64

Dialup. 64
Definition of broadband. 64
DSL 64
Cable modem 66
Broadband connections and electronic security 66
Satellite 67
T1 leased lines 68
Virtual private network 68
VPNs and security devices 69
Internet options 69

Chapter 11   IP Addressing How-To 71
Addressing overview 71
Checking the IP address of a network-connected device 71
PING command 74
Exiting command line screen 74
Command line functions 74
Finding IP information through Windows 74
Summary 77

Chapter 12   IP Addressing Example 79
DSL IP addresses 80
VoIP IP addresses 81
Wi-Fi gateway router IP addressing 82
Computer IP settings 84
Putting it all together 84
Gateway routers and LAN/WAN addressing 85
Summary 86

Chapter 13   Working with the IT Department 87
IT management responsibilities 87
IT management concerns 88
IT and network security 88
How to work with the IT department 89
Going parallel 89
Using the enterprise network 89
Bandwidth controls 90
How much is available? 91
What about the future? 91
Summary 91

Chapter 14   Cabling and Connection Options 93
Standardized structured cabling 93
Cabling as it stands 93
Backbone cabling 94

Horizontal cross-connect 95
Horizontal cabling 95
Network camera installation example 95
    Camera power 96
    Connection to the network 97
    Wait now 97
Parallel networking 97
Benefits of being alone 97
Going parallel-testing 98
Parallel network cameras example—copper 98
Fiber hookup 100
Fiber media converters 100
Viewing options 101
Pre-installation testing of parallel systems 101
10/100 problems 101
Security of telecom rooms 102
Summary 102

Chapter 15   Serial Communications and Ethernet 103
How do these technologies communicate? 103
RS-422 and -485 103
Why change to Ethernet? 104
Ethernet converters/serial servers 105
Head-end communications 106
Serial tunneling 106
Communications integration 107
Access control communication links 107
Summary 108

Chapter 16   Planning a Network Video System Installation 109
The cameras 109
Network camera benefits 109
Network camera concerns or limitations 110
Network cameras—the bottom line 111
Analog camera and network video server benefits 111
Analog camera and network server concerns or limitations 112
Analog camera and network server—the bottom line 113
Monitoring and recording options 113
Summary 114

Chapter 17   Video Compression Technologies 115
Streams of images 115
Basic television 115
Resolution lines 116
Composite video 116

From analog to digital 116
Lossy and lossless compression 117
Decimation 118
Scaling 118
Spatial redundancy 119
Temporal redundancy 119
Further file size reduction 119
JPEG compression 119
MPEG 119
Typical video compression algorithms used in networking 121
Compressing the future 121

Chapter 18   Video Bandwidth Controls for Shared Networks 123
Why bandwidth control is needed 123
Network bandwidth availability and variables 123
Options for bandwidth control 124
    Frames per second (fps) 124
    Image scaling 125
    Scaling, fps, and Internet connections 126
    Compression percentages 126
Compression method selection 126
Variables of bandwidth usage 127
Good, better, best 127
Calculating bandwidth needs 127
Black magic 128
Summary 128

Chapter 19   Video Control and Recording Options 129
Image control 129
Camera movement 130
Image storage, transmission, and recording 131
    In the camera 131
    Email 131
    File transfer protocol (FTP) 132
    Web browser vs. software program control. 133
    Network video recorder (NVR) 133
Compression compatibility issues 133
Multiple image transfer options 134
Sensory overload 134
Summary 134

Chapter 20   Powering Devices 135
Everybody wants some 135
Enterprise power concerns 136
What does a UPS do? 136

Types of UPS 136
   Standby backup offline UPS 136
   Line interactive UPS 137
   On-line UPS 137
Battery backup 137
Intelligent UPS 138
UPS capacity requirements 138
UPS application considerations 138
Remote control and reset 138
Network powering option 138
Power over Ethernet ("PoE") 139
How PoE works 139
PoE power suppliers 139
Intelligent discovery 141
Advantages of PoE 141
Growth of PoE 141
DC pickers 141
PoE and security systems 142

Chapter 21   **Network Security** **143**
Introduction 143
Why is this important? 143
Call the police 143
The high chaparral 144
Who are the bad guys? 144
Inside operators 145
Why hackers might attack 145
How hackers attack systems 146
   Brute force attacks 146
   Computer viruses 147
   Trojan horses 147
   Worms 148
   Denial of service attacks 148
   Logic bombs 148
   Spoofing 149
A "hole" lot of trouble 150
Protecting your network 151
Network protection planning 151
   Physical security equipment 151
   Security of communications lines 152
   Defense against outside attack 152
   Defense against inside attack 153
   Protection of video data 155
   Regular backup of system data 155

Redundant systems/paths 155
Network security is a process, not a goal 156

Chapter 22   DSL Adapter Guided Tour 157
DSL adapter functions 157
Accessing the DSL adapter 157
Password access 158
Firewalls and hosted applications 158
Hosted applications 159
Turn on the host 160
Select the computer 161
Allow individual applications 161
More than one camera? 161
DMZ mode 161
Review the work 162
Getting back to normal 162
Review of DSL adapter setting adjustments 162
DSL adapter review 163

Chapter 23   Wi-Fi Router Guided Tour 165
The router 166
Accessing the device from a PC 166
Wireless settings 166
Turn on/turn off 166
SSID name 167
WEP 167
WEP encryption and key type 167
Disabling SSID transmission 169
WAN settings 169
Host name 170
MAC address 170
MAC cloning 170
DNS addresses 170
LAN settings 171
DHCP settings 171
Static DHCP 172
Advanced settings 172
Port forwarding 172
Filtering options 173
Firewall rules 174
DMZ settings 174
Router review 175
Gimme danger 175
Wi-Fi router review 176

Chapter 24 Ethernet Camera Guided Tour 177

Connecting to an Ethernet camera for initial programming 177

Connecting to the camera using default IP address 177

Connecting to the camera using MAC search 178

Programming network addressing 180

IP addressing—MAC search 180

IP addressing—default IP connection 180

Port address settings 181

Host name 181

BOOTP 181

DHCP 182

DNS 182

Set up and get out 182

Other network settings 182

NTP time settings 182

File transfer protocol settings 184

Ethernet camera tour review 186

Chapter 25 Wi-Fi Camera Guided Tour 187

Why Wi-Fi? 187

Tour camera 187

Programming connections 188

Accessing the programming settings 188

Wi-Fi settings 188

IP address settings 190

Port settings 190

DNS and DDNS settings 191

Set and get out 191

Image settings 191

Email settings 192

Wi-Fi camera review 192

Chapter 26 Wireless Laptop Surveillance 193

Equipment list 194

Camera programming 194

Accessing the camera's programming fields 194

Setting the camera's IP address 195

Programming the camera for ad hoc Wi-Fi mode 195

Programming the laptop for ad hoc mode 195

Security issues 201

"Hey, it's not working!" 201

Recording software 201

Disk space requirement 202

Testing for video storage needs 203

Onsite installation 203
Wi-Fi warning 204
Suggestions 204
Summary 205

Chapter 27   IP Video Server Guided Tour 207
Programming access 207
IP addressing options 208
DDNS functions 208
Image settings 208
Video image options 209
Serial port settings 210
Video server tour summary 210

**Chapter 28   Video Management and Recording Software
Guided Tour** 211
Why use video management software? 211
Software details 211
Program startup 212
Camera connection 212
Video and audio compression 213
Image scaling and cropping 215
Recording options 215
Recorded file encryption 216
Caption selections 217
Motion detection 217
Scheduling 218
Live camera viewing 218
Video management software review 219

Chapter 29   Digital Video Recorders 221
DVR connections and functions 221
Recording options 221
Video motion detection 223
Disk storage options 224
Evidentiary options 224
Viewing options 224
P/T/Z control 224
Networked DVR benefits 225
Connection issues for network DVRs 225
DVRs and hacker security 225
Desktop DVRs—video capture systems 225
Benefits of DVRs 226
Summary 227

Chapter 30	IP Alarm Transmitters	229
	Why transmit alarm signals over networks?	229
	Pick a winner	230
	Receiver compatibility	230
	Losing the signals	231
	Pay to play	231
	Installing an IP alarm transmitter	231
	General programming	231
	IP addressing	232
	Testing of alarm transmissions	232
	Digital dialer backup	232
	Loss of downloading capability	232
	Transmission options	233
	Signal encryption and security issues	233
	Future/now	233
	Practice, practice	233
	Summary	234
Chapter 31	VoIP and Alarm Communications	235
	How VoIP works	235
	Why your customers may switch to VoIP	237
	VoIP and service quality	237
	Installation problems with VoIP	237
	Alarm transmission problems with VoIP	239
	Connection of digital communicators to VoIP	239
	Alternatives to digital communications	240
	Death of the digital dialer	240
	How dealers can protect their business	240
	What's the good news?	241
	It's somebody else's problem	241
	Summary	242
Chapter 32	Tools of the Trade	243
	Ethernet tools	243
	Wi-Fi tools	244
	Product-dependent tools	244
Chapter 33	Testing and troubleshooting	245
	Logical networks	245
	Communications testing sequence	245
	Direct laptop testing	246
	LAN communication testing	247
	WAN/internet communications	247
	Problems with video images	248

No video 249
Slow video 249
Summary 250

Appendix A   Common Networking Terms Glossary 251

Appendix B   Useful Commands for Troubleshooting Networks

and Devices 261

Appendix C   Internet Connection Information Websites 263

Appendix D   Web Pages of Interest 265

Appendix E   Reference Books 267

Index 269

Acknowledgments

The author wishes to thank the following persons for their assistance in bringing this book to life:

Joan Engebretson, wife, mother, and editor

Rob Stritch IV, graphics production

Thomas A. Hoeppner, a special thanks for his technical review prior to publication

Laura Stepanek, *SDM Magazine*

The author would also like to thank the friends he has made during twenty-plus years in the electronic security business: James McIsaac, Jim Mahalak, Bill Russell, Mike Witchie, Cindy Bracey, Tom Thompson, Paul Gulczynski, Joe Kelly, Dean Mason, Jim Hassenplug, Dave Homet, Glen Petit, and the sadly departed Doug McGary.

About the Author

Dave Engebretson is the president of Slayton Solutions Ltd., a Chicago, Illinois company that provides online and instructor-led training in fiber optics and networking of electronic security systems. Entering the alarm industry in 1978, he has designed, sold, and serviced central station, fire alarm, CCTV, burglar alarm, and access control systems. Engebretson is a contributing technical writer for *SDM Magazine,* providing a monthly column on networking as well as the "Kinks & Hints" section.

Introduction

Network connection of electronic security devices opens a vast new horizon for the applications of video cameras, recorders, access control systems, and alarm communications. Today the electronic security industry can provide its clients with the ability to connect, view, record, and control devices from remote locations.

Networking knowledge is critical for today's electronic security firms' very survival. Simply put, if an installation company isn't able to sell and provide networking security options, the competition will, and sales and clients will be lost to them.

While networking provides many new opportunities for our industry, there are also questions about how networks operate, how best to connect security equipment to communications networks, and how to make such connections sure and secure.

This book provides the information needed by today's electronic security installation personnel to effectively interface alarm and monitoring equipment onto Ethernet and Wi-Fi networks, and to allow remote communications with these devices over the Internet. It is the author's intention that by using this manual, electronic security firms will be able to quickly and simply provide the wide spectrum of networking security options now available to their clients.

The information within this guide is written to inform the electronic security professional on the hows, ways, and means of utilizing network communications. It is not my intention to provide a tutorial on the basics of CCTV, access control, or burglar alarms. If the reader requires that knowledge, it is available from a number of qualified sources, such as the National Burglar & Fire Alarm Association's NTS programs.

In my twenty-six years in the electronic security field, I've found that experience is the best teacher. After reading this manual, I strongly suggest that your company obtain a network CCTV camera or interface, and practice connecting the device to a network in your home or office. Much of the knowledge in this manual was developed in this very manner. The mistakes and corrections you make while practicing will reinforce the knowledge you have received from this guide, while reducing or eliminating embarrassment while at your clients' locations.

Networking is a very interesting and exciting technology for our business, where knowledge and creativity can combine to provide new and profitable installation options.

It's a new day in electronic security.

David J. Engebretson
President
Slayton Solutions Ltd./Security Networking Institute
www.SlaytonSolutionsLtd.com

The Beginnings of Networking

The History in brief

Communications networks are everywhere in today's world, connecting computers, telephones, personal digital assistants (PDAs), and carrying personal and business messages around the neighborhood and around the world. Virtually every person and enterprise depends on networks to some degree, whether it is a school, funeral home, business, or a solitary person. Even farmers who live miles from their nearest neighbor use the telephone network to contact the outside world.

Networks are so ubiquitous that they are taken for granted by most users. Few think about the technology involved when clicking on "Google" to search the Internet or pressing "Send" to initiate a cellular telephone call.

Networks transmit two-way information from one entity to another. When using the telephone, one person speaks while the other listens, and back and forth. It is this bi-directional information flow that gives network communications their power, allowing parties or devices to communicate over vast distances, and in the case of cellular technology, even when one or both parties are in motion.

How did these networks develop?

The first fully realized communications network was developed to transmit messages from one station to another over copper cables carrying electrical current. The telegraph used the rapid opening and closing of an electrical circuit from one station to energize and de-energize an electromagnet at the connected station, providing a communications stream of "dots and dashes." Although no longer in use, the concept from the telegraph system of converting alphanumeric symbols (A, B, C, etc.) into a code has continued into the world of computers, where letters, numbers, text, video images, and illustrations are converted into the binary code of zeros and ones.

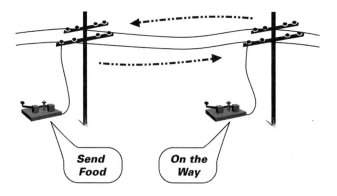

Figure 1-1   Telegraph communication.

Telephone networks

Once developed, the telephone network rapidly replaced the telegraph, providing the ability to transmit voices from one point to another, allowing conversations between two parties. Telephony provided two basic concepts to networking, the first being simultaneous bi-directional communications (also called full duplex) where each party can talk at any time during a session. However, when the two parties talk at the same instant, the information transmitted is often not received. By comparison, the telegraph was limited to half-duplex communications where only one party could transmit at a time; two-way communications required taking turns.

The other conceptual breakthrough in the advancement of the telephone networks was the connection of communications devices to each other by using individual addresses (the phone number) assigned by the network (Figure 1-2). This allowed a single telephone to connect to any other, provided that the user knows the phone number for the person or place he wants to speak with.

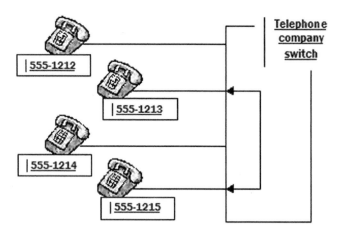

Figure 1-2   Telephone lines.

SECURITY TECHNICIAN'S NOTE: Just as each telephone instrument must have a unique telephone number or extension to differentiate it from others, network devices must have unique addresses to allow them to receive data transmissions that are directed to them.

Mainframe computer networks

Computers were introduced into the business environment in the late 1970s, to allow individuals to access the processing power of a central computer to perform common functions such as data entry, word processing, and other tasks. Computer networking also allowed the use of shared printers, eliminating the expense of providing an individual printer at each workstation.

The initial computer networks installed in businesses consisted of a large "mainframe" or mini-computer, to which were connected terminals, which allowed individuals to input data and access information from the mainframe (Figure 1-3). Connections from the computer to the terminals were usually a special coaxial copper cable run in a "bus" configuration, where all terminals were physically connected to the same length of cable.

These initial computer networks were a tremendous tool for business, allowing data to be accumulated at a central point, the mainframe, and input and accessed from the remote terminals.

As the use of such mainframe-terminal networks increased, the problems and limitations of this type of system became apparent. As the mainframe, terminals, printers, and cables all came from a single vendor, businesses were hostage to a particular vendor for their computer equipment and network needs. And as different vendors used different cable and connection methods, it was very difficult for a

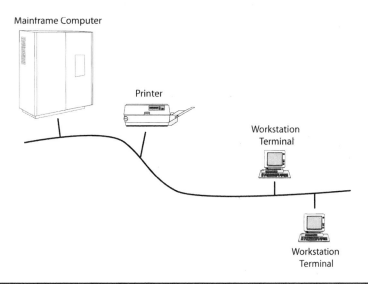

Figure 1-3   Mainframe network.

business to change from one vendor to another. To convert these systems from vendor "A" to vendor "B" would likely require the complete rewiring of the customer's buildings, greatly adding to the cost and complexity for such a conversion.

Cabling system standardization

User groups such as the Electronic Industry Association ("EIA") and the Telecommunications Industry Association ("TIA") came together to address the need for a common cabling standard that would provide a suitable platform for the connection of telephone instruments, computers, printers, and other network devices. Initially released in the late 1980s, and periodically updated, the EIA/TIA 568 standards document details the exact types of cables, connectors, and cross-connection apparatus to be installed within commercial buildings. Further explanation of the details of these standards is included in a later section of this guide.

These standards were rapidly adopted by end users and cabling contractors, providing a uniform method for the connection of networked devices in the workplace.

Personal computers

While the benefits of mainframe computer networks were quickly apparent to the business world, the rise of the Personal Computer (PC) provided the means for individuals to access and use computing power in their daily lives. As the power of microprocessor chips increased, we have seen the development of desktop and laptop computers that can do much more, and cost much less, than the mainframe systems of yesterday. With the increased usage of network protocols such as

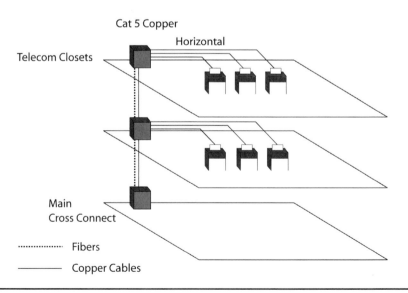

Figure 1-4 Structured cabling.

Ethernet and Token Ring, these powerful PC devices were being connected to business networks, providing users with large scale processing power at their fingertips.

Network protocols

The initial concept behind the development of computer network communications was to allow the sharing of common devices, such as printers, with a number of users. As multiple printers are expensive, it made sense to devise a method where such equipment could be shared.

In the mid-1980s, two networking protocols emerged as the primary contenders, those being "Ethernet" and "Token Ring." These protocols, which you could equate to a spoken language, provided a framework, like grammar, for how information is sent. Over the past decade, Ethernet has emerged as the dominant computer language for network communications. This is because of greater data speeds and easier-to-use cabling methods. Ethernet communications will be covered in great detail in a later section of this guide.

The rise of the Internet

As the usage of networks increased rapidly in the 1980s and 1990s, a need arose to connect various computer networks across the country, and around the world. If an enterprise has office locations in Dallas, Chicago, Los Angeles, and Hong

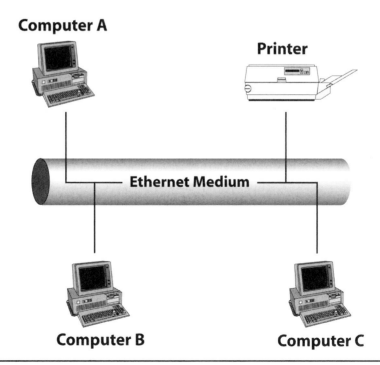

Figure 1-5 Ethernet medium.

Kong, there are many needs for inter-office communications, ranging from simple messaging (e-mail) to complex accounting needs such as billing, receivables, and inventory reporting.

On a parallel track, the military and scientific communities in the United States were interested in the development of a communications network that would be essentially immune to a nuclear attack that could devastate a large area, cutting all communications links into and out of the blast confines. Additionally, they sought to allow easy collaboration between their various locations and the sharing of computer resources. For such a network to continue functioning, information would have to be able to quickly take alternative routes from one point to another. This initial "inter-network" formed by the military/scientific communities is the basis of the Internet we know today. It linked universities, military bases, and other locations onto a network that provided multiple paths for the information to travel. By dividing longer files into shorter "packets," which include a numeric address for the receiving computer, files could be sent along different routes to the receiver, and be re-assembled in their proper order for viewing and usage at the receiving end. Additionally, the use of these small packets allows data that is not properly received to be easily re-sent.

Individuals and businesses soon asked to be included in this network and quickly became accustomed to using the Internet for communications, particularly after the adoption of alternative "name" addresses, such as www.Google.com, that are much easier to remember than their numeric equivalent.

The Internet is now everywhere in our lives, from our business and personal communications to online shopping, and new applications are being developed continuously.

Network history and electronic security

For the security professional, the history of computer networking reveals a mature and vibrant networking industry that combines standardized cabling, computer languages, and network structures into communications networks that are relied on by virtually every business, government, and enterprise. With the increasing production of network devices for the electronic security market, such as Internet Protocol (IP) cameras, video servers, and digital video recorders (DVRs), and IP-addressed alarm panels and transmitters, networking adds new features and benefits for installation companies and their clients.

Of utmost importance to the electronic security industry is the reliability of the products installed. Because of the investment, development, and acceptance by users of these communications networking systems, security installation companies can rely on them to faithfully and accurately perform when transmitting video or security information.

CHAPTER 2

Analog and Digital Communications

Communication methods

To send a voice message, image, or information across telecommunications lines or airwaves requires that the data be converted into a form that can be carried on an electrical current or radio wavelength. The two methods used for this conversion are analog and digital signal transmission.

Analog communications

To transmit information in an analog format over extended distances, the data, image, or voice is converted into an analog waveform, which electrically emulates the input signal. This waveform is then applied to a constant baseband frequency, which is generated by the transmitter. When the signal reaches the receiver, the baseband frequency is filtered out, leaving the analog waveform, which is then converted by the receiver back into its original form of voice or sound, data, or images.

Analog communications have been used since the beginnings of electrical communication, and in the context of security systems are used in common CCTV systems, where the video images are carried over coaxial electrical cables.

Problems with analog communications

Analog communications can be distorted by outside interference, which can result in a poor quality of signal or image. Analog signaling is also prey to attenuation over distance, with signals losing power and quality as they travel through the air or over an electrical cable.

Transmitting in analog form over long distances can require large amounts of electrical power, such as the 50,000-watt AM radio stations that were established in the early 1900s.

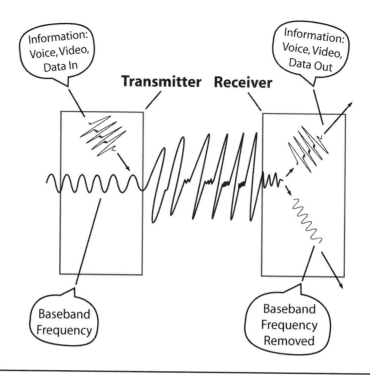

Figure 2-1 Analog communications.

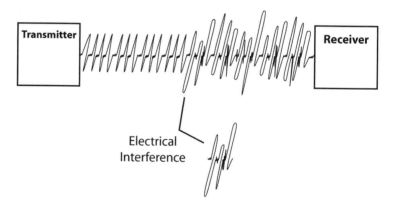

Figure 2-2 Electrical and interference and analog signals.

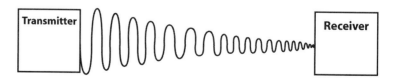

Figure 2-3 Analog distance attenuation

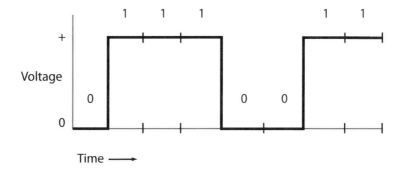

Figure 2-4   Signals to be transmitted are converted into a string of zeros and ones that correspond with specific voltage changes.

Distortion, attenuation, and large power requirements limit the uses and distances over which analog communications are practical. However, analog is perfectly acceptable for short distance use, as is demonstrated by virtually any standard CCTV installation.

Digital communications—zeros and ones

A better way to transmit signals was realized with the development and adoption of digital communications.

As shown in Figure 2-4, signals to be transmitted are first converted into a string of zeros and ones that correspond with specific voltage changes.

These voltage changes are then applied to the "baseband" frequency and sent to the receiver, which removes the baseband wavelength and converts the zeros and ones back into the original data. If the data represented an analog signal, that must be reconverted from digital data back to an analog value.

SECURITY TECHNICIAN'S NOTE: While data is transmitted by changing voltages when sent via electricity, the ones and zeroes are represented by a blinking LED or laser in a fiber optic link.

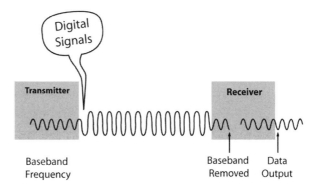

Figure 2-5   Digital communications.

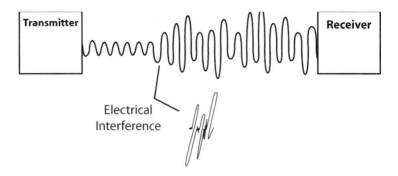

Figure 2-6   Electrical interference and digital signals.

By converting images, voice, or data into a "binary" code consisting of a string of zeros and ones, many of the problems inherent in analog communications can be sidestepped.

Interference

While electrical interference from magnetic or radio frequency sources can cause distortion in an analog signal, it can have less effect on digital communications, as increased "peaks" caused by distortion can still be interpreted by the receiver as zeros and ones.

Attenuation

Digital signaling allows for longer distance transmission potential, as the lessening of the signal over distance can still provide a clear distinction between the ones and zeros.

Signal power

Because the voltages signifying the ones and zeros are of set values, the power needed to transmit digital signals is inherently less than that of analog signals, where the larger peaks of the waveform would require more power to be transmitted.

Summary

With lower power requirements, longer transmission distances and less susceptibility to electrical interference, digital communications are superior to analog, and provide the basis for network communications.

CHAPTER 3

Overview of Ethernet LANs

Evolution

The Ethernet communications language, or "protocol," was initially developed to allow various computer devices to communicate with each other, and to provide for the use of common network equipment, such as printers, by a number of different network users. After development by a research facility, the ways and means of Ethernet were standardized by the IEEE (Institute of Electrical and Electronic Engineers) into a set of standards designated "802.3," as the initial committee was established in the second month of 1980. Additional standards have been adopted over time to codify improvements in the technology.

Overview

Ethernet encompasses a combination of cabling, computer hardware, and software, which together create a community of network communication devices. Properly connected and programmed, Ethernet provides for communications between computers and network devices for file sharing, remote control, in fact every possible aspect of computer usage.

There are two important aspects in the overall view of Ethernet. First is that this technology does not provide for any hierarchy in the local LAN environment; each device on a network or network segment has an equal opportunity to communicate. This means that the company president's computer has the same communication rights on the network as the stock room clerk's.

The second key element of Ethernet is that it is a "software transparent" communications protocol. Ethernet can be used to connect all types of computing devices from main frame computers to "IBM-compatible" desktop PCs to Macintosh machines. And Ethernet is not particular about the type of files or information being transmitted. Ethernet is not "married" to a particular software manufacturer or equipment type; in general, it will work with all suitably equipped computers/network devices, and will transmit any type of data.

Cabling

Devices on an Ethernet network are connected to each other via electrical or fiber optic cabling. As one computer communicates with another the cabled connection often passes through network controlling devices, such as gateways or routers, hubs and switches. Later sections of this guide will describe the typical types of cabling used in Ethernet networks.

Computer hardware

For connection to an Ethernet network, each computer or device must have a Network Interface Card, usually termed a "NIC card." This card contains a standardized connection port, which is usually an RJ-45 eight pin female socket, and "firmware," which is a chipset containing software that performs tasks related to the operation of the Ethernet network. These NIC cards may be removable, such as in a desktop computer, or embedded, as is seen in network-enabled CCTV cameras, DVRs, and some laptop computers. NICs have a MAC address (explained a little later) that acts in part as a unique identifying serial number for the interfaced device.

Firmware

The embedded software (firmware) contained within the NIC converts data to be transmitted into Ethernet "packets," segmenting the overall file into uniform-length data strings to which have been added addressing and error correction coding. The NIC software also provides for the reception of similar packets from other computers, removing the addressing information and forwarding the data. The software within the NIC also controls network timing and collision resolution, which will be covered later in this section.

Individual addressing

To communicate with one another, each computer on an Ethernet network must have an address that is unique to the branch or "segment" of the network to which that computer is connected. This address is commonly called the "LAN IP" address.

Large networks, including the Internet, which is really a network of networks, will contain computers that have the same LAN IP address. To communicate with each other, such computers will use additional address information to individually identify each network device. Just as the street address of a business might be "1290 Elm Street," there may be another firm on the next street whose address is "1290 Pine Street." In this example, each business has the same street number, but the street name identifies each location uniquely. Even further, there may be more than one "1290 Pine Street," but in those cases, the city/state/zip further the identification to make each address unique. The structure of Ethernet provides a number of addressing options and hierarchies, allowing networks to grow quite large, encompassing thousands of computers. Ethernet networks can also be connected to other networks, providing computer connections across the globe if desired.

Nodes

Computers and other Ethernet-enabled devices are commonly called "nodes" on the network.

Segments

Network "segments" are sub-sections of a larger network, typically defined by a cabling element connected to an Ethernet network traffic-controlling device, such as a router or switch. A segment may be connected to one or many network computers or devices. Typically these are parallel-connected devices within one IP address range (more on this later).

How transmission happens

Each device connected to an individual Ethernet network or network segment is constantly "listening" to its network (or segment), and "hears" all transmissions that occur, but only processes transmissions with its own address. Each device ignores packets that are not addressed to it. When one device sends data to another, the receiving device sends a message back to confirm receipt.

What about collisions?

As Ethernet devices can basically transmit at will, the potential exists for two Ethernet devices to transmit at the exact same time, creating garbled data, which is termed a "collision." The solutions to collisions are found in the Ethernet software and in the design and cabling of the network.

When a transmitting computer fails to receive a receipt for a transmission, it will wait a random amount of time, measured in nanoseconds, before it retransmits the previously blocked packet. As a collision requires at least two participants, this waiting and retransmitting of data is performed by both of the colliding transmitting computers.

Although the above scheme helps to maintain the flow of data, real reduction of collisions on large networks is realized through the use of Ethernet "switches," which are hardware and software devices that essentially act as traffic officers, directing packets to the segment to which the receiving computer is connected. By creating a segment for each computer, collisions are greatly reduced.

Ethernet data throughput standards

The initial Ethernet systems were termed "10 Mbps," and had a theoretical data throughput of 10 million bits per second. As communication and cabling technologies developed, "100 Mbps" Ethernet standards were formalized. Today's

NICs will typically be "10/100," with both transmission formats included and available. Such Ethernet devices will "auto-negotiate" between themselves to ascertain at which data speed they will communicate with each other, typically using the highest data speed available.

> **SECURITY TECHNICIAN'S NOTE:** Of the current crop of network-enabled security devices, some do not have the capability for 100 Mbps Ethernet, and only provide transmission at 10 Mbps. Make sure the router or switch to which they are connected can communicate at that speed.

Ethernet goes turbo

The development of higher-speed Ethernet standards continues as end users demand higher throughput of data for larger bandwidth applications such as streaming video and transmission of medical imaging. "Gigabit Ethernet," which provides 1000 Mbps, is already in use by high-end users, and standards are currently being developed to provide 10 and 40 Gigabit Ethernet in the near future.

Ethernet and electronic security

Ethernet provides a very stable framework for the transmission of electronic security signals, whether CCTV images, access control information, control commands, or alarm activations. This technology is reliable, robust, and mature. As Ethernet is the primary networking choice for all sorts of businesses and enterprises, there is wide availability of connectivity for security devices.

CHAPTER 4

Basics of Ethernet

One of the key elements to the success of Ethernet is the concept of "packet" data transmission, where data, images, and even sound files are divided and reformatted into small units, transmitted over the network, and reassembled for use by the receiving computer.

Why data is reformatted for transmission

It is certainly possible to transmit a file in a complete stream from start to finish from one computer to another. This type of transmission is often used when a system only transmits in one direction, from one device to another. Think about a typical analog CCTV setup, where one camera is connected to a single monitor. The camera sends its images in one uninterrupted stream, and the pictures are displayed on the monitor.

The situation becomes complicated when multiple communication devices, such as computers, are connected to a common cable or "segment," and each device needs the capability to speak with all of the others. If files were sent from one computer to another in one uninterrupted stream, all of the other devices would be unable to communicate during that time period.

Another problem with the transmission of files as complete streams is if an error in the data occurs. If the file transmission is corrupted by interference, the file must be sent again in its entirety to complete the transfer from one computer to another.

Because of these issues, Ethernet divides files to be transmitted into packets, which provide addressing, sequencing, and other information, that allows the receiving network device to properly reassemble the file. These packets are also referred to as "frames."

Ethernet packet format

The 802.3 standards specify how Ethernet packets are assembled for transmission.

Let's look at Figure 4-1:

Destination Address	Source Address	Sequence Code	Data	Frame Check Sequence

Figure 4–1   Ethernet Transmission Packet.

The "Destination Address" is the address to which the packets are going. For obvious reasons, this is placed at the beginning of the packet stream. The "Source Address" is needed, so that the destination computer knows where to send receipts for packets and, if necessary, requests to resend packets that may have been garbled during transmission. The "Sequence Code" is the number of the packet within the overall file being sent. If a file has 567 packets in its transmission, the sequence code says, for example, that this is packet #324 of 567 total. This allows the data to be reassembled in the correct sequence, once it is all successfully received.

The "Data" section is where the actual content is located. This data field can be a minimum of 46 bytes up to maximum of 1500 bytes, with a byte being eight bits (ones and zeros).

The "Frame Check Sequence" provides an error correction mechanism to ensure that the complete packet arrived without error. Simply put, as the receiving computer accepts a packet, it counts the total number of bytes in the packet. The Frame Check Sequence provides a number, indicating the total number of bytes that should have been received. If the receiving computer's byte total doesn't match the Frame Check number, the receiving computer will request a retransmission of that particular packet, referencing the Sequence Code (packet number).

Packet routing

On large networks such as the Internet, packets may travel by various routes to get from one computer to another. As the packets travel, "routers" and "switches" read the destination address of the packet, and forward it along to its final destination. Routers and switches direct the packets to the fastest path available at the time the individual packets are received. If one path becomes busy with other traffic, the router may send a packet along another route.

Device addressing

To facilitate communications, each device on an Ethernet network must be properly and uniquely addressed. Just as the telephone companies must issue phone numbers that are not duplicated, the manager of an Ethernet network must carefully program the computers and other devices on the network so that data can be transmitted and received from one device to another. What follows is a general discussion of the role of addressing in Ethernet networks; later will come detailed information on various methods by which devices are addressed.

MAC addressing

The "Media Access Control" or MAC address provides a product serial number, or "physical address," which is used to identify a particular device on a network. This is a "hard" address, which is implanted into the "firmware" of the product at the factory. In most cases this address cannot be changed by the user, although later we will see situations where devices can have their MAC address changed.

Every device that can be connected to an Ethernet network will have a unique MAC address. Similarly, any network-enabled electronic security device will also be coded with its distinctive MAC. For example, the MAC address of a particular Panasonic MV-NM100 network camera is

00-60-45-0D-56-EC

By combining letters and numbers in this format, literally billions of MAC addresses are possible.

SECURITY TECHNICIAN'S NOTE: The first three groups of a MAC address are a manufacturer's distinct code; the second three groups are the serial number of a unique device. So all Ethernet devices produced by vendor "C" will have the same first three groups in the MAC address.

Think of the MAC address as the "Vehicle Identification Number" (VIN) for your automobile or truck. Usually found on a stamped metal plate, your car's VIN is viewable through the windshield from the outside, typically on the driver's side. This is a unique number identifying your particular vehicle, and no other car has the same number as yours.

The MAC address is used on Ethernet networks to uniquely identify individual devices. When a file is transmitted, the data packets contain the MAC address of the recipient, and the assigned IP address.

IP addressing

The address format that we are most concerned with for networking applications is the "Internet Protocol," or IP address. This numeric code is settable and changeable, allowing the network administrator or installer the ability to modify settings to achieve proper communications.

Current IP addresses are formatted into four groups of numbers, with a maximum of three digits in each group, called "octets." An octet is 8 bits. There are no letters used in current IP addresses. A typical IP address for a computer on a network might be:

192.168.1.23

Any number in an IP octet must be between 1 and 255. So an address such as:

192.168.343.2

is not valid.

IP addresses are considered to be "logical" addresses, and they define the type of network to which the device is connected and whether it is a local network (LAN), wide area network (WAN), or the Internet.

Network classes and the Internet

IP addresses define the type or "class" of network to which a device is connected. There are three classes of networks—A, B, and C.

Class A networks

The largest international networks and ISP providers use Class A networks. To illustrate, let's look at the IP address of www.Yahoo.com, a large Internet service provider:

216.109.118.75

Now let's examine the IP address of www.yahoomail.com, which provides email access for those with Yahoo accounts:

216.109.127.29

The "216" in the first octet is common between the addresses, identifying them as members of the same Class A network.

Over sixteen million individual computers can be connected to a single Class A network, all having unique IP addresses that start with the same octet, which in Yahoo's case is "216."

Class B networks

When the first two octets of two different IP addresses are identical, they are both part of the same Class B network.

Let's look at the IP address of www.yahoomail.com again:

216.109.127.29

Notice that this address matches the first two octets of the www.yahoo.com address, those being "216" and "109." So while Yahoo has a unique first octet, "216," it also has subset networks that are Class B.

Each Class B network can have up to 65,000 individually addressed computers connected to it. Typical Class B network users are larger companies and universities.

Class C networks

When the first three octets of two different addresses are identical, they are both part of the same Class C network. Because of the restriction of numbers over 255, a class C network can have a maximum of 255 computers or network devices connected to it.

Here's the IP address of www.sbcYahoo.com, another address used to communicate with Yahoo:

216.109.127.30

Notice that this address matches the first three octets of the www.yahoomail.com address.

Typically a Class C network with be a local area network (LAN), confined within one or a few buildings that are within close proximity to each other. As there can only be a maximum of 254 network devices connected, this limits the size of the network. However, it is a simple matter to connect many Class C networks together, allowing communications from one network to another.

Class A and B IP addresses are issued by the Internet Corporation for Assigned Names and Numbers ("ICANN").

Reuse of IP addresses

If connected to separate networks, devices often have identical IP addresses. The common form of IP addressing for LAN devices is in the format of "192.168.0.XXX," or "192.168.1.XXX," with the "Xs" providing up to 254 distinct addresses for devices. ICANN has reserved 192.168.XXX.XXX addresses for this purpose, making sure not to assign them to any particular entity. While there may be literally millions of network devices with the address "192.168.1.1," because they are each on unique LANs, there is no conflict in addressing or communications. Later we will explore how like-addressed network devices can communicate using Network Address Translation, or "NAT." Think of this like being able to have the same phone number in different area codes.

Default IP addresses

Network cameras, video servers, and other electronic security network devices are generally shipped with a pre-programmed "Default" IP address. This address will typically be something like:

192.168.1.20

or

192.168.0.10

This default address will most likely have to be changed by the installer to enable communications to and from that particular device.

Addresses and communication

To allow communications between devices on a local network, the IP addresses of each connected computer must be programmed the same for the first three octets of the address string (and also for the subnet mask, which is explained below). For example, if you want to use your laptop to program a network camera that has a default address of 192.168.1.15, you will need to check and/or change the IP address of the laptop to 192.168.1.XXX, with the X's being any number from 0 to 255, excluding "15," which is the default address of the camera.

Think of the IP address as the license plate number on your car. Your plate number is unique within your state, but others may have the same plate number issued by another state. And your license plate number can change if you are issued a new one, or splurge for a "vanity" plate. The MAC address, which is equivalent to the VIN of your vehicle, doesn't change just because a new license plate number was assigned to the car.

Why two types of addresses—IP and MAC?

Networks function on a number of levels, and the two addresses of a device are used by various network devices to direct or route data traffic from one computer to another. Later the uses for these addresses will be discussed in detail.

Another reason for the two addressing schemes is to provide for the changing of equipment on the network without having to totally reconfigure the addresses of connected devices. For example, a network-connected printer can be upgraded, and the new one can be programmed with the same IP address as the unit that is being removed. This way the logical address stays the same, and all relevant devices can communicate with the new printer without network reconfiguration.

Subnet masking

Subnet masks are programmed into a device along with the IP address and are identical for every device on a LAN. For local area networks, 255.255.255.0 will be the typical subnet masking address.

Each device on the LAN will only accept messages from or send messages to addresses that begin with exactly the same first three octets. The subnet mask, in effect, tells other devices to look only at the last octet when reading a device's IP address.

Summary

Each Ethernet network device will have two addresses, a MAC (physical) and an IP (logical) address. MAC addresses are hard-coded into a device or product at the factory, and typically are never changed. IP addresses are settable and changeable, allowing networks to be configured and changed.

Ethernet Devices and Components

Ethernet networks are composed of various devices, which provide specific functions in the transmission of data packets from one computer to another. These device types are common to all Ethernet networks, large or small, and also to the Internet. Electronic security technicians must understand what these devices are, and their common features, to best integrate security devices with existing network equipment when performing an installation.

Ethernet devices

The following details the different types of devices used on an Ethernet network and the function(s) that the device performs.

NODE—A node is an addressed network device that can transmit or receive data packets. Common nodes are desktop or wired laptop computers and printers. Devices such as network cameras, video servers, alarm interfaces, and audio interfaces are also considered to be "nodes," as they are addressable and can send and receive data.

NIC—(pronounced "nick") Acronym for a Network Interface Card. Typically a circuit card installed in a desktop computer, laptop, or printer, the NIC provides connection to the Ethernet network, typically via a female RJ-45 eight-pin socket. The NIC contains firmware that performs the packetizing functions needed to transmit data over the network. The NIC contains the MAC address for identification purposes. NICs may either be a separate card in the chassis of the computer, or may be built-in, as they are in network cameras and video servers.

MEDIUM or MEDIA—Copper or fiber optic cabling that connects the devices on an Ethernet network. Most of today's networks use a copper cable, with four pairs of 24-gauge individually insulated conductors under one jacket. A single section of cabling is called a "segment," which may have one or more Ethernet devices connected to it. The types and ways of cabling Ethernet networks are covered in a later section of this guide.

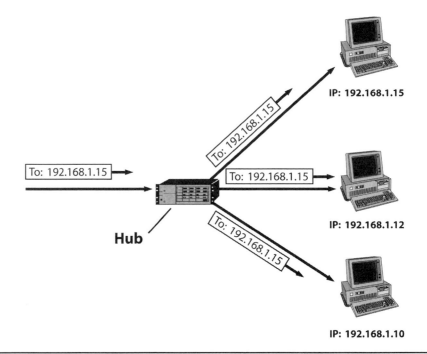

Figure 5-1 Hub communication.

HUB—A hub is a retransmission device that connects a number of Ethernet segments together. When data packets are presented at one connection on the hub, it rebroadcasts them to all other connected segments. Because all of the segments connected to a particular hub are electrically connected, the available bandwidth for data transmission is shared between all connected devices. Hubs are the least expensive way to expand an Ethernet network.

SWITCH—A switch is similar to a hub, in that a number of different Ethernet segments are connected to a single device. Switches contain circuitry and software that provide the switch with the ability to "learn" and remember the devices that are connected to it. When data packets arrive at the switch, the switch retransmits the packets to the specific node to which they are addressed. Because of this directing capability, devices connected to the switch do not share their bandwidth capacity with other nodes on that particular switch. Switches can be as inexpensive as hubs for small networks, and provide greater throughput.

GATEWAY or ROUTER—A router connects two different types of networks together, such as an enterprise's LAN and the Internet. Routers are programmable, and can provide firewall protection and Network Address Translation ("NAT"), and can shield internal addresses from outside networks. Routers have two MAC addresses, and two LAN/WAN IP addresses, one for each network to which they are connected.

Data transmissions between local computers are kept within the LAN by the router, which examines the "subnet masking" address, and sees where the data should go.

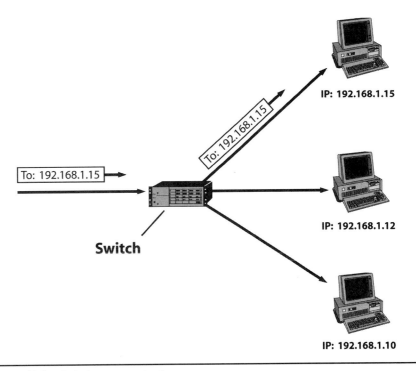

Figure 5-2 Switch communications.

When an internal computer requests a Web page on the Internet, or communicates with any device on the Internet, the router replaces the transmitting address (the local computer's) with its own WAN address. When the answer to the request reaches the router, the router re-addresses the incoming packets so that they reach the requesting internal computer. This protects the enterprise's computers from outside penetration.

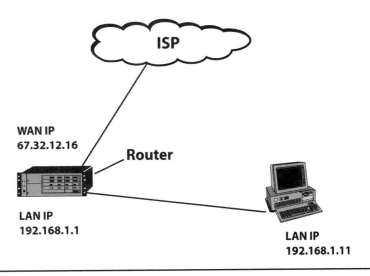

Figure 5-2 Router communication.

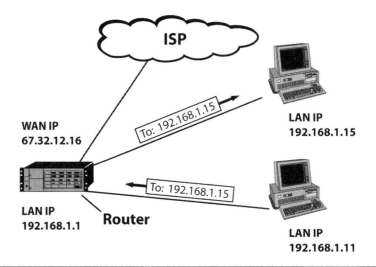

Figure 5-4   Local router communications.

Even inexpensive routers can have many sophisticated programming features, for example allowing Internet access only at certain times during the business day.

Although a single computer can be connected to the Internet via a DSL, cable modem, or T1 line, the addition of a router provides needed protection against outside intruders. A very common type of router combines Wi-Fi capabilities with four or five Ethernet connections. This combination router is often used in homes or small businesses, allowing multiple wired and wireless computers to share a single Ethernet connection.

BRIDGE—A bridge connects two LAN networks together, typically between two buildings. One bridge device is connected to each network, only passing data packets between them that are destined for the other network. The bridge devices identify each other by their respective MAC addresses. The media connecting the bridge devices can be fiber optic, copper, or wireless.

ADAPTER—An adapter connects a LAN network (or individual computer) to the Internet. Typical adapters provide connections to DSL and cable Internet providers.

GATEWAY—This is another term for a router. When inputting the IP address into a device being connected to a LAN that uses a router, the "Default Gateway" is the LAN address of the router.

FIREWALL—A firewall is a hardware and/or software device that protects nodes from unwanted intrusion by other computers on the network and/or Internet. Firewalls check the leading data packets entering a network and either allow or disallow the traffic based on pre-set tolerances. Firewalls have adjustable settings to allow online gaming, connection of net cameras, and other hosting applications that may be viewed over the network and/or the Internet. *Firewalls will often block video transmissions, and must be disabled or reprogrammed when connecting network-enabled cameras or video servers.*

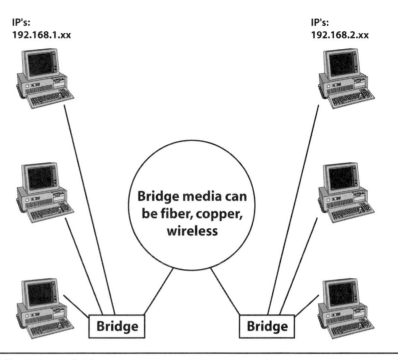

IP's:
192.168.1.xx

IP's:
192.168.2.xx

Bridge media can
be fiber, copper,
wireless

Bridge

Bridge

Figure 5-5   Ethernet bridge

SERVER—a "server" is a computer that holds specific applications or programs
and allows authorized clients to access them. Typically a server will require a "username" and a password before allowing such access. Network-enabled IP cameras,
DVRs, and analog-to-network video interfaces act as servers, providing their video
images to "clients" which communicate with them. Simply put, servers give, and
clients take.

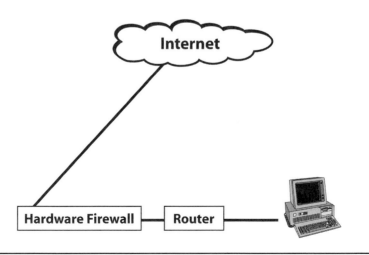

Internet

Hardware Firewall — Router

Figure 5-6   Firewall.

CLIENT—a client is the other side of the server/client relationship. Clients are allowed access to all or certain applications, data, or programs that reside on or are provided by the server.

MODEM—A modem is a device that reconverts digital transmission into analog for transmission, typically over standard phone lines. The word "modem" comes from compressing the words "Modulator-Demodulator."

Proxy server

Many medium and large size networks will use "proxy servers" to help speed and monitor WAN and Internet communications. The proxy server is a gateway, connecting outside networks to the internal LAN.

Configured as either a software program or a separate server computer, the proxy server provides three important functions. The first is Network Address Translation ("NAT"). Packets from the LAN that are directed to the WAN or Internet are stripped of their local address, and the WAN address of the proxy server is substituted. When data is returned from the Internet, the proxy server re-addresses the packets to the original requesting LAN IP address.

The second function of a proxy server is "caching." As an example, consider an enterprise network that allows users Internet access. Users regularly click to the "Google" web page to look up information. The proxy server will locally store the "Google" web page, and provide it to the local users when requested instead of reconnecting to Google over the Internet for each request. When users type in their requested information and click "Google Search," the proxy server then sends that information over the Internet. Caching provides faster communications for commonly requested WAN or Internet web pages.

Logging Internet usage is the third primary function of a proxy server. All packets that pass between the Internet and the LAN can be recorded, providing a log of which computers have access to the LAN from the Internet, and which Internet web pages local users are accessing. This is a powerful management feature, providing a recorded trail of Internet/LAN communications.

> **SECURITY TECHNICIAN'S NOTE:** It is important to understand that although the NAT and logging functions of a proxy server happen very quickly, each and every data packet must be processed, and these actions add additional time delays to communications. Video image transmissions, because of the large number of data packets involved, will be slowed if passing through a proxy server network gateway. If remote DVR, video server, or network camera functions must pass through a proxy server to be viewed over a WAN or the Internet, frames per second rates will be reduced.

Summary

Different devices are used to control and route data traffic to, from, and through an Ethernet network. Hubs and switches provide paths for data within a LAN, while routers control the connections of LANs to larger networks such as the Internet.

CHAPTER 6

Wire and Cable

To best utilize the power and capabilities of Ethernet data communications, the connections between devices need to provide large bandwidth capabilities, and be uniform in how devices are installed onto the network. This uniformity is important to all parties involved in building, installing, maintaining, and using Ethernet networks.

Initial cabling methods

When Ethernet began to gain popularity, the cabling systems initially used evolved from the proprietary cabling schemes that were used for terminal-mainframe systems such as IBM's. These proprietary systems used a coaxial cable, to which desktop PC computers were connected via a penetrating pin that pierced the outer jacket of the coax, and made an electrical connection with the conductive core of the cable. These connections were commonly termed "vampire taps."

Just like their proprietary cousins, the first Ethernet systems were connected in a "bus" configuration, where all devices were physically connected to the same length or "segment" of cable.

Rethinking the concept

A number of issues caused a rethinking of the use of these coaxial cable bus configurations for Ethernet networks. Having a large number of communicating devices connected to a single cable increases the likelihood of data collisions and the failure of a single device or cable failure could cripple the entire network.

In general, workers need a network computer and a telephone to perform their job from a workstation. Therefore the development of a new cabling standard was also propelled by the need for the standardization of telephone cabling. Just as end users did not want to be held hostage by proprietary computer cabling, they also did not want to limit their options for upgrading or changing their telephone systems.

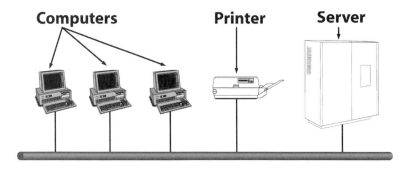

Figure 6-1 Bus topology. In a "bus" configuration, all network devices are connected to single cable.

The EIA/TIA (Electronics Industry Association/Telecommunications Industry Association) formed committees in the late 1980s to develop new standards for communication cabling within buildings. These written standards, called EIA/TIA 568, have been embraced by end users, installers, and product manufacturers to the extent that almost all cabling, connectors, and Ethernet devices are built to conform to these standards.

Star configuration

To reduce the potential for data collisions and eliminate the servicing issues of having multiple nodes on the same cable, the 568 standards call for cabling to be installed in a "star" configuration. Each computer or connected device is connected to its own "segment" of cable, which are then run to centralized connection bays, typically called "cross-connects."

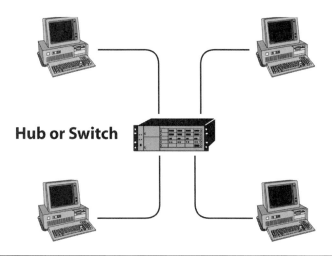

Figure 6-2 In the "Star" configuration, each computer or network device is connected on its own cable to a central hub, switch or gateway.

By installing Ethernet switches at the cross-connects data collisions are greatly reduced. As each device is on its own "segment," troubleshooting of problem nodes, cables, or connections is greatly simplified.

568 standards overview

In simple terms, the EIA/TIA standards specify the following parameters for communication cabling:

1. Type of cable
2. Type of connectors
3. Maximum distances of cable runs
4. Testing methods

Let's examine each of these subjects, and see how they can affect the installation of network-enabled electronic security devices.

Type of cable   The 568 standards specify the use of twisted pair copper cabling or fiber optic links between connection points. The copper cable consists of four pairs of conductors, with each conductor individually jacketed, and with each pair twisted together to reduce the effects of electromagnetic interference (EMI) and radio frequency interference (RFI). These four pairs are held under an overall outer jacket. In most installations within the US, this copper cabling does not include an overall shielding element. Common terms for this form of copper cabling are Unshielded Twisted Pair ("UTP"), "Category 5," "Cat 5," and other variations on the "Cat" theme.

Figure 6-3   UTP consists of four pairs of insulated twisted copper conductors under a single jacket.

Fiber optic links are specified as multimode (62.5 or 50 micron core) or single-mode (9-micron core) fiber cables.

Types of connectors   For copper connections, a male and female format called the "RJ-45" is specified by the standards. This connector set provides eight electrical connection points, one for each of the conductors in a four-pair Cat 5 cable.

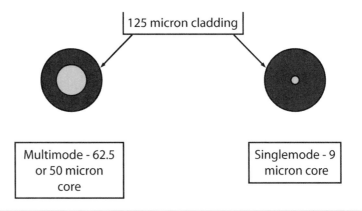

Figure 6-4 The type of fiber is defined by the size of the care, either single mode or multimode.

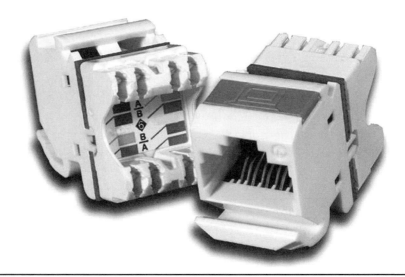

Figure 6-5 Network connections are made using standardized connectors.

What is also specified is where each conductor is connected within the jack, both male and female. These connection schemes are often called the "pin out" configuration. There are two accepted formats for the copper connection sequence. These are called 568A and 568B. Figures 6-6 and 6-7 show 568A and 568B connections.

Note that the standards provide "same to same" connections, so that when the connector is plugged into the socket the green conductors are connected together, the green/white is connected to the green/white, etc.

It is very important that the copper connection pin-out sequence be uniform throughout a particular installation. In the great majority of networks, 568B is

568 A Jack Termination

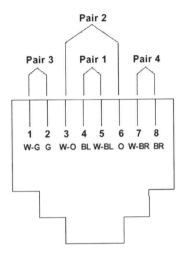

568 B Jack Termination

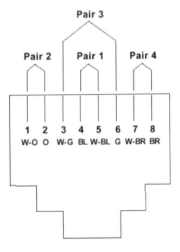

Figure 6-6 and 6-7   568B is the most popularly used connection.

used. If there is a question, check with the IT manager to confirm the standard used for a particular system.

The fiber optic connectors specified by the 568 standards are the "SC" type. Although this connector type is called for in the standards, the most popular type of connector used in fiber is the "ST."

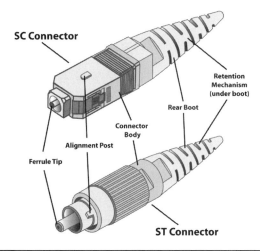

Figure 6-8 Common fiber connectors. Either SC or ST connector styles are typically used for fiber terminations.

The construction and uses of fiber optic links will be discussed in an upcoming section of this guide.

Maximum distances of cable runs

A key element of the EIA/TIA 568 standards is the strict limitation on the maximum distance of cable runs. Starting at the workstation or wall outlet, the maximum cabling distance to the cross-connection location is 90 meters, or 295 feet. The maximum lengths of the "patch cords" that connect network devices to each other through the wall outlet and the cross-connection are a total of 10 meters, or roughly 30 feet. These cable runs from the wall outlet to the cross-connection point are often called "horizontal" cabling, as they are usually installed within one floor of a building.

SECURITY TECHNICIAN'S NOTE: It is good practice to limit the installation distances of new UTP communications links for the connection of network cameras or servers to 100 meters or less.

"Backbone" cabling, which provides communications between cross-connection bays, is governed by a different set of specifications in the EIA/TIA 568. The distance limitations of copper and fiber backbone cables are listed in Table 6-1.

TABLE 6-1   Backbone Distances

Media Type	Maximum Backbone Distance
UTP Copper	1640 ft. (500 m)
62.5/125 Multimode Fiber	5575 ft. (1700 m)
50/125 Multimode Fiber	5575 ft. (1700 m)
9/125 Singlemode Fiber	8855 ft. (2700 m)

Testing methods

To verify the quality of installation, structured cabling systems are typically tested for a variety of properties, such as return loss, proper connections, lengths of links, and other technical standards. Such testing is performed using sophisticated two-piece "Cat 5" or "Cat 6" tester sets, and the results are generally provided to the client upon the completion of the cabling project.

Ethernet copper connections

10 and 100 Mbps Ethernet (sometimes called "Fast Ethernet") use two of the four pairs in a UTP cable to transmit and receive data.

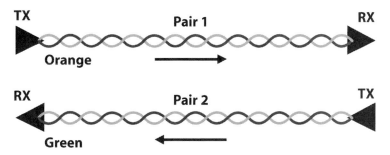

Figure 6-9 Two Pair 10/100 Ethernet. The orange and white/orange pair is used to transmit data, while the green and white/green receive.

To reach the higher data throughput required, Gigabit Ethernet uses all four pairs for transmission.

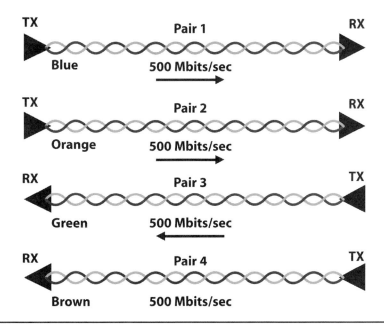

Figure 6-10 Four pair gigabit Ethernet.

This increases the potential for interference, as there are two additional data streams under the single jacket containing the conductive pairs. Any external interference can be picked up by any of the four active pairs, and the possibility of crosstalk between pairs within a cable increases.

UTP patch cords and jumpers

Computers and other network devices are typically connected to RJ-45 wall sockets using "patch cords" or "jumpers." These cables are pre-made with Cat 5/5e/6 UTP cordage, and factory installed male connector ends. The patch cord type selected should be of equal grade to the horizontal cabling connected to the wall outlet; i.e. Cat 5e patch cords should be used if Cat 5e cabling is installed.

> **SECURITY TECHNICIAN'S NOTE:** Patch cords are the "weakest link" in cabling systems. As they are exposed to use and abuse by client personnel plugging and unplugging devices, rolling over them with chairs and furniture, etc. patch cords can degrade in performance. Use new patch cords to connect electronic security equipment.

Crossover cables

If a PC or laptop is to be directly connected to another Ethernet device, without first going through a hub or switch, a "crossover" patch cable must be used.

In a crossover cable, the orange conductor is crossed to the green, and the white/orange is crossed to the white/green. What is "transmit" on one end is "receive" on the other.

> **SECURITY TECHNICIAN'S NOTE:** Crossover cables are often needed to perform the initial programming of networked electronic security devices. Most manufacturers do not supply a crossover cable with their products, so the technician will need to provide one. To help prevent confusion with other types of cables, use crossover jumpers that are red in color.

Copper cable performance enhancements—all those "Cats"

To provide an adequate electrical pathway for higher speed Ethernet communications, cabling and connector manufacturers have greatly improved the data transmission capacities of their products over the years. As manufacturers have improved products, end users have requested that such improvements be codified by the EIA/TIA.

The initial 568 specifications for computer network copper cable and connectors called for "Category 5," which provided an end-to-end bandwidth minimum of 100 MHz, which can be roughly equated to the Mbps performance capability of that cable and connector combination.

"Category 5e" and "Category 6" provide increased bandwidth performance, which enables these cable & connector combinations to successfully transmit high-speed Gigabit Ethernet.

Copper cable installation and performance

As data speeds increase, how a UTP cable is installed and terminated takes on much greater importance. While 10 Mbps Ethernet may be able to pass through a poorly installed cable and/or connector, 100 and Gigabit Ethernet communications may slow significantly over the same cable.

Summary

Typical cabling systems within commercial buildings usually conform to accepted industry standards. It's important for security contractors to fully understand the particular cabling system within a client's building, and to test all pre-installed cables and connectors that will be used to provide communication pathways for electronic security devices.

CHAPTER 7

Fiber Optics

Fiber optic links provide the largest bandwidth and furthest distances for the transmission of Ethernet data in networks. To utilize fiber optic links properly, the security technician must understand how fiber works, the different types of fiber that are typically installed, termination, testing, and fiber-to-electrical data conversion devices.

Why fiber is used in networking

In structured cabling systems, fiber is generally used to provide connections between cross-connection rooms, or between buildings in a campus environment. Fiber is used in these instances because it has much larger bandwidth capabilities than copper, is immune to RFI and EMI interference, and can transmit signals for distances much greater than copper cable can provide. Where an analog CCTV camera can be connected to a monitor using coaxial cable out to a distance of perhaps 1000 feet, the same camera can be connected to a fiber, using converters, and the video signal can be transmitted successfully over many miles.

To a much lesser extent fiber is installed to the work area or desktop, typically for high-bandwidth users such as hospitals and advertising agencies. These users require large bandwidth capability to rapidly transmit and receive graphic files.

How fiber works

Electrical data signals, such as Ethernet, are converted into blinking light pulses, which are injected into one end of a fiber link. These light pulses are at specific frequencies that are invisible to the human eye. The light pulses travel the length of the fiber link, are received and then converted back into their electrical state.

Because the fiber is glass, and the signals within are light pulses, this technology is completely immune to outside electrical interference. The reconstituted signal coming from the receiver is exactly the same as what was input on the transmission side.

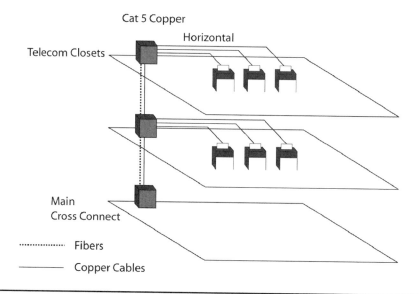

Figure 7-1 Fiber is often used as the "backbone" for large bandwidth communications between computer rooms or buildings.

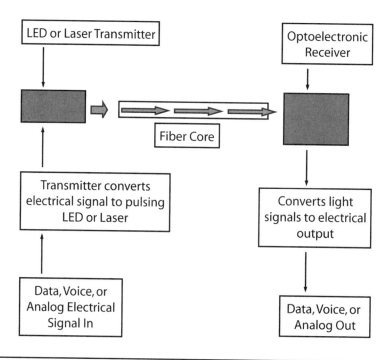

Figure 7-2 How fiber works.

Types of fiber and their uses

There are two common types of fiber that are used in network installations, multimode and singlemode.

Multimode fiber has a core size of 62.5 or 50 microns (a micron is one-millionth of a meter). Multimode fiber is used for link distances of less than four miles.

Singlemode fiber has a core size of 9 microns. It is used for link distances longer than four miles. Singlemode has greater bandwidth potential than multimode, so it is also used to provide high-bandwidth connections, regardless of distance.

If the fiber is already installed, it can be difficult to determine which type of fiber it is, as it can be difficult to determine the difference between the two types of fiber without specialized tools and testing equipment. Usually the fiber is labeled on the outside jacket, indicating the quantity of fibers and their type, singlemode or multimode.

> **SECURITY TECHNICIAN'S NOTE:** The lasers used to transmit over singlemode (and some multimode) fiber links are dangerous to the human eye. This is particularly hazardous, as the frequencies used are outside the visual spectrum; a technician can be exposing his eyes to powerful lasers without actually seeing a light. Technicians should be particularly careful when using a microscope to examine connector ends, as the microscope will amplify any laser light coming through the fiber link. Be certain that transmitters are disconnected before viewing the end of a fiber connector.

Fiber connections

A fiber "link" consists of a connector on one end, a length of fiber, and a connector on the other end. Fiber connectors provide a flat and smooth surface that allows the maximum amount of light to pass from one connector to another, or from a connector to a network device. Fiber connectors are typically one of two types, either the SC or the ST.

Connector installation requires specific tools and procedures to ensure that the finished connector can pass light through it. Fiber connector installation is a specialized skill requiring extensive training and practice.

Fiber testing

A simple flashlight often is the only tester a security technician needs to verify that a fiber link is ready for use. By pressing the connector tip from one end of a fiber link onto a flashlight, visible light will be carried to the other end of the link for distances up to two miles. If the light is visible at the other end, the fiber link is not internally broken. If the light shines through, the link being tested is most likely usable for analog CCTV and other low-bandwidth applications such as access control and alarm signaling. Flashlight testing provides a quick and simple "no go" test; if the light doesn't pass through, the fiber link or most likely its connectors are defective or broken.

Using an "optical loss test set" provides a quantitative measurement of a fiber link's quality. Measuring in dB (decibel) of loss, the optical loss test set is a two-piece tester, including a light source and a measuring meter. The optical loss measurement can be compared to industry standards for a specific type and length of fiber, providing a reliable indicator of the quality of the fiber cable, connectors, and any splices included in the tested link. Optical loss testing is needed when the fiber link is going to be used over long distances and/or requires large bandwidth capability. A fiber that passes the flashlight test may be OK for a simple 10 Mbps Ethernet connection, but if the same fiber has high optical loss due to installation problems, it may not be capable of passing 100 or 1000 Mbps data communications.

While the flashlight indicates that the fiber link isn't broken, and the optical loss test set measures fiber link quality, neither tester can inform a technician as to the location of a fiber break, bad connector, or other problem. The Optical Time Domain Reflectometer (OTDR) provides a visual readout of a fiber link, indicating the location of problems within a few feet. OTDRs inject a specific wavelength laser beam into the fiber link being tested, and receive reflected signals from connectors, splices, and the length of fiber itself. These signals are amplified and interpreted by the OTDR's circuitry, and a "trace" is presented on the OTDR's screen, providing detailed location information for the technician's viewing and analysis. OTDRs are invaluable for the servicing of long-distance fiber optic links. People using these devices should be properly trained and should use the devices carefully to avoid misinterpretation of results.

Electrical-to-fiber converters

As CCTV cameras, access control systems, and alarm systems operate using electrical communications, these signals must be converted into light signals to be transmitted over a fiber link, and reconverted into electrical signals at the receiving end.

Analog transmission on fiber

Both digital and/or analog signals can be transmitted over fiber optic links. In the case of digital transmission, the transmitter's light source turns "on" or "off," indicating the ones and zeros of binary data signals. Analog signals are emulated by the transmitter's light source brightening or dimming.

Standard CCTV cameras can be connected to fiber optic links using analog interfacing device sets, which convert the incoming electrical video signal into an analog signal that is transmitted over the fiber link and reconverted at the receiving end. These device sets are usually fiber-type specific (multimode or singlemode), and can provide multiple video signals and bi-directional signaling. These capabilities enable pan/tilt/zoom operations to be controlled and alarms to be transmitted over one or two fiber links.

Some vendors offer "digital" fiber transmitter/receiver sets that convert the incoming analog CCTV video signal into a digital stream, which is passed over the fiber link to the corresponding receiver. While such product sets may increase the quality of the signal received, it is important to note that the received output of such a "digital" device set is an analog CCTV signal, not a packetized Ethernet data stream.

Analog fiber transmission sets work well when connecting existing CCTV cameras to remote buildings for monitoring and control. However, these devices should be thought of as "coax extensions." They don't provide any of the benefits of truly networked video cameras, which can provide networked video imaging, intercom, alarm inputs and outputs, and access control communications.

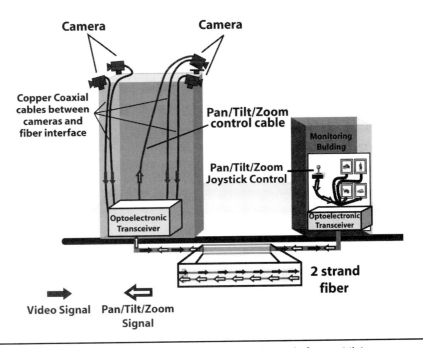

Figure 7-3 A single fiber can carry video and control signals for pan/tilt/zoom CCTV cameras.

Ethernet media converters

When using fiber optic links to connect Ethernet-enabled video cameras, servers, and the like to the larger network, common Ethernet "media converter" sets provide the means of converting electrical Ethernet data into digital light pulses which are reconverted into electrical data signals at the other end.

It is important to remember that for devices to function properly on an Ethernet network they must be capable of both receiving and sending data packets. So an Ethernet media converter is by necessity a two-way device which transmits data over a fiber link while simultaneously converting received data from a separate fiber link back into its electrical component.

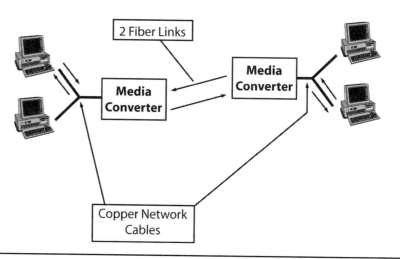

Figure 7-4 Media converters change electrical data signals into fiber optic light pluses, extending network connections.

For Ethernet connections, the typical converter set will use two fibers, one to transmit and one to receive. If only a single fiber link is available, more expensive media converter sets can be deployed. These device sets use bi-directional light signals with two distinct frequencies providing the necessary two-way Ethernet communication.

One attractive feature of media converters is their simplicity during installation. Typically there is no software to install or manipulate, and few setting or connection options. Just plug the RJ-45 jack from the network into the converter, hook up the fiber link(s), apply power, and the converters should be communicating with each other, passing Ethernet data back and forth.

Selection of media converters

When selecting electrical-to-fiber media converters for use in networked security installations it's important to know the fiber link type (multimode or singlemode), number of fibers available (two are better than one), connector style (ST or SC), and the bandwidth requirements of the cameras or video servers that are to be connected to the fiber for transmission. Many of today's network-enabled video servers and cameras are 10 Mbps Ethernet. This can cause a problem when using common Ethernet media converters, which may only function at 100 Mbps.

When proceeding with a new installation, it is a simple process to "bench test" the suite of selected equipment prior to the actual installation. By pre-testing the network video devices with media converters using short fiber optic link "jumpers," the technician can be assured that the proper data flow from the security devices onto the network can be readily achieved.

When to use media converters for security applications

Because electrical-to-fiber media converters can be used in any network application and not just for electronic security connections, such devices are produced in much greater quantity than electronic security-specific devices such as the specific analog interfacing equipment previously discussed.

Whether to use media converters or analog interfacing devices when connecting to fiber links depends on a number of factors. The vital question is whether network features such as viewing/recording on network PCs and/or over the Internet is desired. If so, existing analog cameras can be connected to network video servers to provide these functions. On the other hand, if a simple connection of video and/or alarm devices to a monitoring station in a nearby building is needed and networked security device benefits aren't required, using analog interfacing devices is the most straightforward approach.

Powering devices

As with all electronic security components, it's important to consider what will happen if and when primary AC power fails. In the case of the analog interfaces and media converters detailed above, a power outage will interrupt any data transmissions until the power is restored. Optimally such devices would be connected to an Uninterruptible Power Supply (UPS) that provides power for a short time period after the failure of the primary power source.

Computer-based devices such as media converters do not take kindly to "dirty" power input, and may possibly "lock up," or stop operating, after a power surge, electrical noise spike, or restoral of power after an outage. It may be necessary to manually reset such "locked up" devices by turning them "off" and then "on" again, or removing the power cord and reconnecting after a few seconds.

On-site security personnel should be trained in the procedures for resetting network devices in the case of power anomalies, and written instructions for resetting electronic security system devices should be provided to your clients.

Summary

Fiber provides the highest bandwidth capacity of any cabling medium, while being completely immune to the effects of RFI and EMI. There are millions of unused "dark" fiber segments within existing network cabling plants, and these provide excellent pathways for the connection of network security devices. During the sales and system planning process, security dealers should always ask about the availability of unused fiber links. If such links are available, they should be properly tested for functionality before the start of a system installation.

CHAPTER 8

Wireless LANs

While networks wired with twisted pair copper cabling and/or fiber optic links provide high bandwidth and functionality, there are many situations where pre-installed network cabling is not available. Wireless networks provide a high quality connection for networked devices. They also offer the flexibility and ease of use that has made wireless technology attractive for a variety of non-network applications.

Wireless networks can be particularly useful for electronic security applications, as wireless network cameras can be quickly and easily installed and programmed without pulling new communications cable.

The history lesson

The need for a quality wireless network that would allow the ready connection of laptop computers and other devices has been apparent since the inception of networking. Initially, a small number of manufacturers responded to this need, generally with proprietary product sets that were often quite expensive and cumbersome to install, as they required a "site license" from the FCC (Federal Communications Commission) for the specific frequencies used by the particular wireless system. Because of these issues, few of these proprietary systems were deployed.

The concepts of Wi-Fi

The common wireless network systems in use today fall under the general description of "Wi-Fi," which stands for "Wireless Fidelity." Wi-Fi is a combination of a number of concepts, which together have produced an easily installed, secure, low cost, and functional set of wireless network devices. The building blocks of Wi-Fi are:

Use of Shared Frequencies

The FCC determines who and what can use specific portions of the radio frequency spectrum. Companies such as broadcast AM and FM radio stations, television stations, and others have obtained licenses from the FCC for the specific frequencies at which they operate. These licenses include the frequencies to be used, maximum power outputs, and other issues. The FCC has also reserved certain blocks of frequencies as ISM, which stands for Industrial, Scientific, and Medical. These frequency blocks are characterized by low-transmission power requirements and limited range, and do not require site licenses. Users can transmit and receive on these frequencies using off-the-shelf devices. Many common devices use the ISM frequencies, including cordless telephones and some wireless alarm transmitters.

Wi-Fi devices operate in one of two ISM bands, either 2.4 or 5 GHz. Using these shared frequencies allows Wi-Fi equipment to be easily built and installed.

Emulation of Ethernet

As wired Ethernet is the predominant networking technology in common use, it made sense to develop wireless network devices that closely follow the same methods of addressing. Wi-Fi devices are addressed in a very similar fashion to Ethernet, using the same format and inputs. Just as each address on an Ethernet network must be unique, so must the related wireless devices have addresses that are unique to all other devices on the network.

Standardization

The IEEE (Institute of Electrical and Electronic Engineers) has established standards for Wi-Fi components, which are closely followed by manufacturers of wireless networking equipment. These standards are labeled as the "802.11" standards, and they specify the exact functions of Wi-Fi components. Here is a general description of the common 802.11 standards:

802.11b- Freq. 2.4 GHz- 11 Mbps
802.11g- Freq. 2.4 GHz- 54 Mbps
802.11a- Freq. 5 GHz- 54 Mbps

The above descriptions indicate the frequency band at which devices meeting a particular standard operate, and the maximum bandwidth of that particular standard. Later we will learn that the process of packetizing and securing wireless data will reduce the data throughput capabilities of Wi-Fi networks to less than the maximums stated above.

SECURITY TECHNICIAN'S NOTE: There are 11 sub-channels used in 802.11b/g Wi-Fi communications, from 2.412 to 2.462 GHz. Most access points and Wi-Fi routers come defaulted from the factory on Channel 6; as many users don't change the default value in their device, Channel 6 can become congested in areas where there are multiple Wi-Fi networks in operation. Selecting another channel besides Channel 6 will often provide higher throughput and better performance.

Product compatibility

One of the basic concepts in the formation of the Wi-Fi industry was to allow many manufacturers entry into this market in the hope of driving technical innovations and reducing component costs. If a number of vendors are making Wi-Fi equipment, how can users be assured that a device from vendor "A" will inter-operate with a device from vendor "B"?

The Wireless Ethernet Compatibility Alliance (WECA) is an industry organization that tests wireless network products for their compatibility with other vendors' devices. When a product has passed this test, the vendor can display the "Wi-Fi" logo on the product, its packaging, and advertising. This allows users to purchase Wi-Fi devices with confidence, knowing that the products should inter-operate with products from other vendors now and in the future.

Data security

As Wi-Fi transmits data through the airwaves, security from eavesdropping or hacking is a very important issue. Wi-Fi products include the option for data "encryption," where a user-selectable security code is programmed into each device in a specific wireless network. This security code triggers a mathematical function called an "algorithm," that in effect scrambles the data being transmitted over the air, and unscrambles it when the data is received. This encryption feature is called WEP, for Wired Equivalent Privacy. It is important to note that most Wi-Fi products are shipped with the WEP function disabled, and it must be turned on by the user to provide data security.

Wi-Fi components

In general all Wi-Fi devices must have the ability to transmit and receive data, provide for individual addressing, and allow for the programming of encryption security. Wi-Fi technology is incorporated into a few typical network devices, which are defined by how they connect to the network, either wired or wireless, and whether they provide for data transmissions from a single device or allow multiple devices to communicate with them.

Access points

To provide Wi-Fi connectivity in an office or building, a device called an Access Point (AP) is connected to the existing Ethernet network via a twisted pair UTP cable with an RJ-45 connector. The AP can be programmed to provide network access for a large number of individual Wi-Fi-equipped computers or devices. Multiple APs can be installed to extend coverage and increase the number of potential users on the system. Bandwidth is shared by all Wi-Fi devices that are currently communicating with a specific AP, so data throughput can slow down as the number of current users increases.

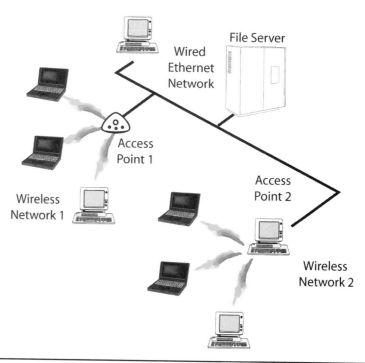

Figure 8-1 Wi-Fi network.

Wi-Fi laptops

One of the most common ways that Wi-Fi is used is by a person equipped with a laptop computer that has either an internal or plug-in Wi-Fi transceiver card. Once the laptop has been addressed by the user (static IP) or received a "leased" temporary address from the AP (dynamic IP), the user can access the network and use any network components or features that are allowed by the system administrator, such as printers and Internet access.

Wi-Fi routers and switches

Commonly used in a small office or home environment, Wi-Fi "routers" provide multiple hard-wire connector sockets for RJ-45 connections, and also the functions of a Wi-Fi access point. These devices can provide a variety of programming options, including firewall settings, Network Address Translation (NAT), and other features that the electronic security technician must set to allow devices such as video servers and network cameras to be viewed from outside the LAN via an Internet connection.

Ad hoc mode

Wi-Fi devices can be programmed to communicate directly with each other, eliminating the need for a separately installed AP or Wi-Fi router. This is called the "Ad Hoc" mode, and can be used to allow two or more Wi-Fi equipped laptops to share files and other resources. By using the Ad Hoc mode with a Wi-Fi network camera, a laptop computer can be utilized as a DVR (digital video recorder) for temporary or covert surveillance. Be aware that not all Wi-Fi network cameras can be programmed in the Ad Hoc mode; check with the manufacturer before planning an installation of this nature.

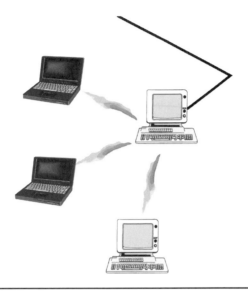

Figure 8-2 Wi-Fi enabled computers can communicate with each other without the need for a separate AP or Wi-Fi router.

Wi-Fi coverage

As with all low-powered wireless transmission systems, the area of coverage of a Wi-Fi access point, router, or Ad Hoc-enabled device can vary widely from one installation to the next. Metal cabinets, pipes, foil-backed insulation, and other structural factors can greatly limit the coverage of a Wi-Fi system. The only sure way to test for Wi-Fi coverage for a particular installation is to program the device to be installed, such as a network camera, so that it will communicate with the Wi-Fi AP or router, and test the connectivity of the device when it is placed in its intended location. Various after-market antennas are available that can be connected to either the AP/router or the remote device to improve reception. In general, such specialty antennas must be installed at the AP or Wi-Fi router, as the "field" device often isn't equipped with a connection port for the addition of an external antenna.

Excessive coverage is another area of concern. If a Wi-Fi system provides coverage beyond the walls of a home or building, unauthorized users or hackers can sit in a parking lot next to the building and attempt to access the network. Special antennas can be installed that provide specific areas of coverage to reduce or eliminate the ability of outsiders to access the network.

Wi-Fi security concerns

The security of Wi-Fi communications is a greatly debated topic among IT (Information Technology) professionals. Many are concerned that a Wi-Fi network provides an open door for unauthorized users to "hack" their networks. Network professionals are also wary of "rogue" access points that have been installed by company workers without the approval of network management.

There are a variety of programming options in Wi-Fi devices that can greatly increase security, such as limiting the number of computers that can use the system, disabling the transmission of the network's name, and limiting the Internet access of users by date, time, or authorization level.

The key to successful Wi-Fi security is to enable WEP encryption, and change the WEP password or code on a regular basis. Change all of the default names and addresses that are programmed into a Wi-Fi device by the manufacturer, and use uncommon passwords and usernames, including numbers, letters, and symbols such as %, #, and &.

To successfully "hack" into a Wi-Fi network that has its WEP encryption enabled, the hacker uses commonly available "sniffer" software installed in a Wi-Fi- equipped laptop, and attempts to obtain the WEP algorithm by sampling literally millions of packets as they pass through the air. The more users on a Wi-Fi network, the more data will be transmitted. So busy Wi-Fi networks are easier to hack into, and system administrators should be more diligent about the periodic changing of the WEP encryption coding.

Although the entry of unauthorized users is a major concern when using Wi-Fi networks, remember that access to the network doesn't necessarily open the entire network and its Internet connections to potential abuse. Which files or folders are available for others to use (called "sharing" in Windows), which computers can access the Internet, and when such access is allowed are all controllable options for the system administrator.

Here is a checklist of available options that can be employed to tighten the security of a Wi-Fi system:

1. Always change the default usernames and passwords on Wi-Fi access points and routers. Use unusual characters and password combinations as mentioned above.

2. In the Wi-Fi device program, disable the broadcast of the "SSID" name. This is the Service Set Identification, which is the name that is given to a particular Wi-Fi network. If this feature is not disabled, the Wi-Fi access point or router will be constantly broadcasting its name, essentially asking other Wi-Fi computers to try to connect to it.

3. Enable WEP encryption, and change the algorithm code regularly. If hacking is suspected, change the code immediately. Remember that once this code is changed in the AP or router, all authorized computers and devices such as Wi-Fi network cameras will need to be reprogrammed for the new coding, or they will be unable to communicate.

4. Review the programming options of the particular AP or router being used. Many of these options can provide additional security measures to make hacking the system more difficult and/or less attractive. Some options may include MAC address specification, which limits access to the network to computers and devices with specific MAC addresses, and limiting Internet access by specific users so that only certain users can access the Internet, and only on specific days at specific times of the day.

5. Use a Wi-Fi-equipped laptop programmed to the network being used as a tester of Wi-Fi coverage. Power up the laptop, get connected to the network, and carry it out to the parking lot and around the building. If the network in question provides coverage into parking areas or outside the building, replace the existing antennas on the AP or router with directional antennas that will provide coverage within the building but not out into public areas.

6. Use a hardware or software firewall package, and scan all network computers for viruses at least once per month.

7. Keep firewall software and the OS (Operating System, such as Windows XP) program current by periodically checking their websites for software patches and virus fixes.

Just as no electronic security system can guarantee a client that it won't be burglarized, the implementation of Wi-Fi security measures doesn't ensure that hacking won't take place. Using the security tools available in Wi-Fi devices is akin to the "target hardening" or "detection in depth" provided by a burglar alarm system that includes magnetic contacts on the doors and internal motion detectors, which can sense an intruder even if he has entered an area without activating the door contact. There are literally millions of Wi-Fi networks operational in the world today; in fact there are four networks that my laptop can reach from my desk in Chicago, Illinois. Remember that hackers, like burglars, will attack targets that are easy to penetrate. "Hardening" a Wi-Fi network makes the hacker's task much more difficult, and he may well choose to attack an easier target.

Bandwidth realities

The bandwidth values of 802.11 Wi-Fi networks as stated in the standards will not be provided by a typical Wi-Fi installation. Interference, WEP encryption, reflections, and other radio frequency-related issues combine to reduce the average data throughput of these systems. A typical "real world" 802.11b system may provide 2-4 Mbps, which is perfectly adequate for the transmission of text files and small graphic images. At the time of this writing, (mid-2004), all wireless electronic security devices—such as network cameras—on the market use the 802.11b technology, probably because the cost of the transmission chipsets is quite inexpensive. Using this technology will limit the fps (Frames Per Second) rates that can be achieved when transmitting video from such a device. Through the use of video compression, fps rates of Wi-Fi cameras can reach 8–12 fps, which can provide an adequate image for viewing. Later in this manual we will examine video compression and its effects on the file sizes of streaming video and images.

Typical Wi-Fi uses for electronic security

The Wi-Fi products currently available for electronic security use are network cameras. Here are some applications for using Wi-Fi technology in security installations:

1. Temporary CCTV Viewing and Recording—Without the need to pull new cables, Wi-Fi equipped network cameras can be quickly and easily installed to provide temporary or permanent viewing of indoor or outdoor scenes. It is a simple matter to program a Wi-Fi camera to communicate directly to a Wi-Fi enabled laptop using the "ad hoc" mode. The laptop can record video (and from some cameras, corresponding audio), providing NVR (Network Video Recorder) functionality for a covert surveillance installation. (Such an installation is detailed in Chapter 26.)

2. Residential CCTV—Using Wi-Fi cameras for residential situations removes the problem of wiring in finished homes, while providing maximum flexibility for both the installing company and the end user. With a functional Wi-Fi router or access point in place, cameras can be placed around the home, inside and out, to provide local viewing on the homeowner's computer(s) or via remote Internet access. By using different network addresses and port numbers, each wireless camera can be individually accessed by authorized users on the network. With a high-speed Internet connection, these devices can be accessed from any Internet connected computer in the world. Software included with the cameras or purchased separately can record all images onto a remote or local network computer for later review.

3. Moveable cameras—Some Wi-Fi cameras such as the SOHO and Veo products can be set on a shelf or desktop as a stand-alone device. Using these cameras residentially, caregivers can move the wireless camera to the back porch, baby's room, or other area, and use their network computer or laptop in another room to watch their children. Remember that the computer viewing the Wi-Fi images can also be wireless. Plug the camera into a wall socket in one room; watch it from any Wi-Fi laptop within the coverage area of the network.

   Commercial uses for moveable cameras may include temporary installations, industrial processing and management, remote monitoring of "problem" areas in production, and other uses that have yet to be conceived.

4. Wi-Fi Bridges—When installing a discrete network for electronic security use that requires connection of network segments in separate buildings, a pair of Wi-Fi access points can be programmed to function as a bridge connecting the two network segments together. By using directional antennas and specific programming, Wi-Fi access points can connect two or more

network segments, which can themselves be wired Ethernet, Wi-Fi, or a combination of the two. This is a powerful tool for installing companies, as it can eliminate the expensive installation of outdoor cabling to connect one building to another.

5. Portable monitoring stations—A Wi-Fi equipped laptop with the proper pre-installed software can be used as a portable monitoring station for a networked video and/or security system. In the event of the forced evacuation or criminal takeover of the primary guard station, such a laptop can be used to monitor and control the video and security systems. Successful use of a Wi-Fi computer in this scenario will require careful preparation and testing to determine the extent of wireless coverage, so that emergency operators know how and where to set up the Wi-Fi monitoring station.

The future of Wi-Fi electronic security

Once an installing company has mastered the programming and functions of Wi-Fi cameras, the sale and installation of these devices should quickly become a highly profitable addition to the security company's product line. Residentially, every client with children and/or concerns about what happens at their home while they're out will want to view the situation from the office or laptop computer. Commercially, the possibilities for incorporating Wi-Fi cameras and devices are only limited by the installation company's or user's imagination.

Summary

The vast proliferation of Wi-Fi networks provides many opportunities for security companies to install network cameras and other devices with reduced labor time and costs. Wireless network cameras can be portable, allowing end-users (or security dealers) to customize video surveillance to meet a specific situation or environment. It's important to remember that wireless communication is inherently more vulnerable to interception or intrusion by unauthorized users; wireless networks used for security signal and/or video transmissions need to be configured using encryption to reduce security risks.

CHAPTER 9

IP Addressing Technologies

Because there are many types of equipment and networks, there are a variety of technologies used to provide IP addressing for network devices. This chapter will examine and explain these addressing options and also the terms and concepts related to IP and Internet addressing.

Host

Any computer or IP addressed device on a network can be called a "host." A host can be called by its IP address or can be given a name, such as "NetCam 2" or "Dave's Computer."

Server

Servers are computers that store and allow access to specific programs or services. Larger companies may have email, database, or other types of servers for specific functions. Other users on the network can be allowed access to some or all of a certain server's information or programs. Servers are used to concentrate data and programs, providing easier maintenance and control of computer services.

Static IP

"Static" IP assigns a specific IP address to a specific device. These addresses typically won't be changed very often, hence the concept of "static." New devices added to the network will need to have a compatible address programmed, and the network administrator will need to keep careful track of which devices have which particular addresses to prevent assigning the same address to multiple devices.

Static IP is often used in smaller networks that do not often have devices added to them.

Dynamic host configuration protocol—DHCP

In many networks, there are temporary users such as outside salespeople or corporate management who wish to connect their laptop to the LAN when they visit an office location. Dynamic Host Configuration Protocol (DHCP) is an addressing program, provided automatically by a network router or gateway, that provides temporary IP addresses to network computers or devices that request them.

To set up DHCP, network administrators can select a number of options, based on the sophistication of their DHCP-equipped router or network gateway. Options include how many DHCP addresses will be issued at one time, what the range of addresses will be, how long each address will be active ("DHCP Lease"), and other security options.

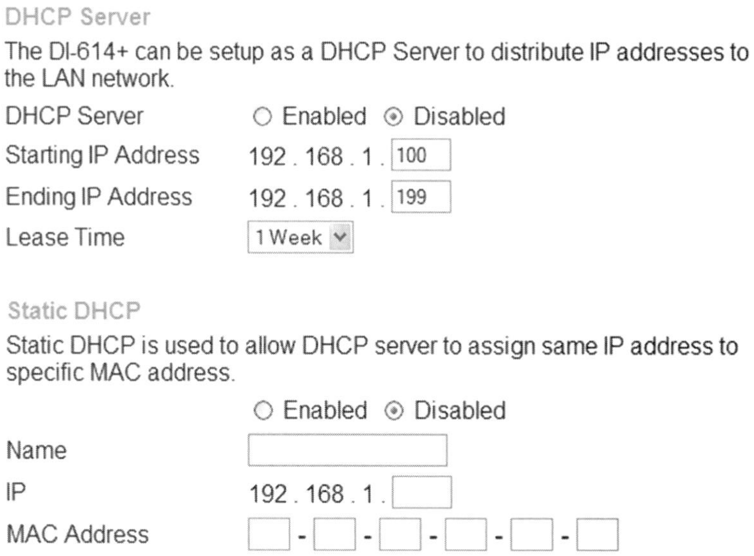

DHCP Server

The DI-614+ can be setup as a DHCP Server to distribute IP addresses to the LAN network.

DHCP Server   ○ Enabled ⊙ Disabled

Starting IP Address   192 . 168 . 1 . 100

Ending IP Address   192 . 168 . 1 . 199

Lease Time   1 Week ▾

Static DHCP

Static DHCP is used to allow DHCP server to assign same IP address to specific MAC address.

                     ○ Enabled ⊙ Disabled

Name

IP   192 . 168 . 1 .

MAC Address   - - - - -

Figure 9-1 This DHCP setting screen from a Wi-Fi routes shows various options that can be manipulated by the system manager.

Using DHCP greatly reduces connection difficulties for temporary users and for networks that are often changing and/or adding computers. When DHCP is enabled, users can simply set their computers to accept an IP address, without contacting the system administrator. Windows users can simply select DHCP in the "Internet Protocol (TCP/IP) Properties" window associated with the particular NIC or Wi-Fi connection to be used, and the network does the rest. This eliminates potential inputting errors by casual users who may not be familiar with the proper setting of IP addresses, subnet masks, etc.

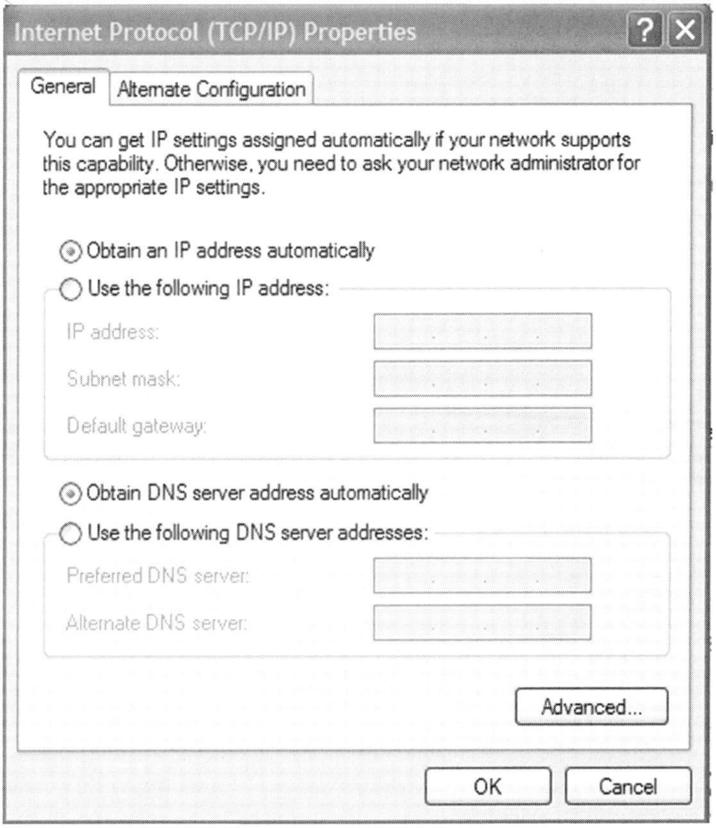

Figure 9-2 Using these setting in windows XP allows the selected NIC to receive a DHCP IP address from the network.

DHCP is quite often used by Internet Service Providers (ISPs) to provide addresses for DSL and cable modem adapters.

Static vs. DHCP in electronic security applications

In general, electronic security devices such as network cameras and video/alarm servers should be programmed with static IP addresses. One of the key benefits of these devices is the ability to access them from outside their LAN via an Internet connection. If this capability is to be available, the IP security device must have a static IP address so that the router knows where to send requests for access or information from outside the LAN. This will be fully explored below in the section on Network Address Translation (NAT). Also, some security devices on the market cannot be programmed to accept DHCP addressing.

IP alarm signal transmitters, on the other hand, can work well with DHCP enabled, as their periodic "polling" signals will provide updated addressing information to the central station receiver.

Networks can operate with a combination of static and DHCP addressing. For example, the "fixed" desktop computers connected to a LAN might be static, while a few DHCP addresses are enabled to accommodate the Wi-Fi-enabled laptops of visiting personnel.

Both Ethernet and Wi-Fi routers can be selected to work with DHCP, static IP addresses, or both, based on programming selections.

SECURITY TECHNICIAN'S NOTE: While the best situation for network DVRs, cameras, and video security devices is to have static IP addresses on a local network, some IT managers are reluctant to provide static addresses for security equipment. One way around this problem is to ask the IT manager for a "DHCP Reservation" for the device(s) to be addressed. If the network's DHCP server has this capability, it can be programmed to provide a specific LAN IP address each time a specific MAC-addressed device requests a network address. Programming the MAC address of the device(s) into the server will effectively provide a "static" or unchanging IP address for the security device, while the IT department can use their DHCP server to administer and monitor the addresses on their network.

Another issue to note is that while some network-enabled security devices can be programmed to ask for a DHCP address, some don't have this capability. Check the programming features of the products selected to be sure.

Domain name server—DNS

Addresses on the Internet are actually numeric, such as the address for the White House's web site, www.whitehouse.gov, which is 64.164.108.149. To make the Internet easier to use, Domain Name Servers (DNS) are computers to which common name web addresses, typically starting with www, are transmitted. DNS servers retain a log of known IP addresses, and cross-reference the common name web addresses to their actual numeric address. Large networks may deploy their own DNS server, while users of ISP services will generally use a DNS server(s) as specified by the Internet network provider.

Which DNS server a particular computer utilizes for address conversions is selected by the user. If DHCP addressing is selected in a PC with Windows software, it can also be programmed to automatically accept DNS server addresses from the DHCP server. Or the DNS server(s) can be set manually, while using DHCP for the LAN network address.

A DNS server must be employed to allow the conversion of www.whatever.com type addresses to their numeric IP equivalent. If no DNS server is selected or programmed, that computer will be able to reach other network devices using the four octet IP address, but not by using www.-type addresses.

Dynamic domain name server—DDNS

A Dynamic Domain Name Server (DDNS) provides a method for tracking the changes of DHCP addresses, usually for computers and/or networks that are connected to the Internet with DSL or cable adapters. A small server program in one of the computers connected to the DSL or cable modem tracks the current IP address and transmits that address to a web site that has a static IP address. Users register on that web site to have an IP address tracked, and can go to that web site to find the current IP address of a specific location. This service can typically be used at no charge.

This is a critical service for electronic security applications, where a network camera, video server, or other device must be accessible over the Internet, and it is connected to a DSL adapter or cable adapter. Most of these Internet connection devices are using DHCP to receive their address from the ISP, as this type of addressing is the least expensive. If the client wants or needs to connect to the camera or security network device at the remote location, and the dynamic IP address has been changed by the ISP, the client cannot remotely access the security equipment over the Internet without having a pre-programmed DDNS service in place. If the IP address has changed, the client can visit the selected DDNS service, enter a password, and receive the current IP address for that location.

DDNS services can also be set up to provide a unique host address that can be automatically forwarded to a client's IP address. So a unique address, such as **http://acmestore.camera,** can be used to access the camera in a retail location. When this IP address is called over the Internet, it reaches the DDNS service, which connects the request to the current IP address of the specified client system.

An alternative to using a DDNS service for IP tracking is for the remote subscriber to telephone someone at the location, and have that person type **www.whatismyip.com** into the web browser of a computer connected to the network in question. The IP address of the network will be displayed on the user's computer screen, and can be relayed to the remote viewer. This only works if there is a person onsite to request the IP address of the network at the time that it is needed.

Ports & network address translation—NAT

As described in a previous section, gateway routers provide a method of masking the existence of LAN-connected computers and devices to the Internet. For example, when a computer on a LAN makes a request for a web page the router uses its own WAN address to make the request, masking the address of the individual computer. This "firewall" function helps protect individual computers from outside hacking or compromise.

However, there are instances where it is desirable for LAN-connected devices to be accessible from the Internet, as is the case for a network camera or video server that the client wishes to view from a remote location. How can the gateway router provide for such remote access without broadcasting the internal IP addresses of LAN-connected devices?

Network Address Translation (NAT) can be programmed to provide a path for incoming Internet communications through the gateway router to designated LAN-connected devices.

NAT uses TCP/IP "ports," which can be imagined as the different channels on a cable or satellite television system. In this example, a single cable may be connected to a television, yet it provides potentially hundreds of separate channels for viewing. What channels are available for your viewing depends on which channels have been "turned on" by the cable provider for your location and cable interface box.

TCP/IP ports function in a similar manner. There are thousands of available ports, with numbers ranging from 0 to 65536. Ports numbered 1024 and below are most commonly used. Some port usage is standardized, with web traffic coming to port 80 and email to port 25.

Ports are used to direct communications traffic through the gateway router as seen in Figure 9-3.

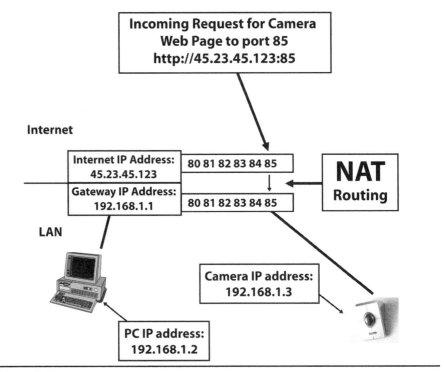

Figure 9-3 NAT Routing.

In this example, the camera has been programmed to respond to communications on port 85. When an Internet user outside of the LAN wants to connect to the camera, he inputs the WAN IP address of the gateway router, 45.23.45.123, adding the suffix ": 85." The router has been programmed to translate Internet communication requests addressed to port 85 to the LAN IP address 192.168.1.3. If there were multiple cameras, video servers, or other devices on this LAN that are to allow viewing or control from the Internet, each such device must be programmed to respond to unique port numbers (86, 87, 88, etc.). The NAT would also need to be properly set in the gateway router to allow such communications to pass through. Once outside users have reached the camera or other device, they will be challenged to provide a proper username and password before the local device will allow communication.

Once the port address has been programmed or changed in a device, typically that device will only respond to requests from other local LAN computers that also include the specific port address. In the above example, if the camera had been left at its default port, 80, it would respond to a communications request addressed "192.168.1.3" from another computer on the LAN. If the port address has been changed, as in the example above, other computers on the LAN would have to use "192.168.1.3:85" to reach the camera.

Summary

Internet Protocol (IP) addressing is the basic technique that security technicians must master to successfully program network-enabled security devices onto a LAN. IP addresses can be issued either statically or dynamically, and both types of addresses can reside on the same network at the same time. IP addresses on a specific LAN must be unique, with no duplications. IP addresses are also used for Internet communications, where DNS servers resolve URL "language" addresses into their underlying numeric IP addresses.

TCP/IP software ports are used to establish communications between network devices. NAT provides a method where a specific port number can be forwarded through a router to a specific IP address, allowing outside authorized users to reach a DVR, network camera, server, or other device from outside the local network.

Internet WAN Connections and Services

The Internet is now providing the electronic conduit for all sorts of businesses, services, and applications used by millions on a daily basis. From soldiers on the other side of the world to businesses, schools, and churches, a great and growing number of individuals and enterprises are connected to the Internet. These connections provide email communications, business transactions, and information for millions of users.

There are various ways in which Internet access can be provided to a user. When planning and connecting electronic security devices to the Internet, it's important to understand the technologies at work with each different type of communication, and the potential impact on electronic security devices and communications.

What is the Internet?

The Internet is a conglomeration of thousands of individual computer networks, large and small, all connected together to allow certain types of communications. The Internet "backbone" consists of very powerful network routers, connected via fiber optic links that direct TCP/IP data packets between themselves, providing a pathway between the sending computer and the receiver.

Although there are international organizations that control some aspects of the Internet, such as issuing domain names such as www.securitynetworkinginstitute. com, it is important to understand that the Internet is not "owned" by any one person, company, or government. And while there are computer language protocols and laws regulating the conduct of those using the Internet, literally anyone can connect any type of computer to the Internet and attempt communications. Whether such communications are helpful or harmful to other Internet users can only be determined after the fact, which is why network security, firewalls, and other protection technologies are so important.

Internet service providers

First, a basic understanding of the hierarchy of the Internet from the connection perspective is needed. While very large enterprises such as the Federal government may have direct connections to the Internet backbone, most companies and residences receive their connection from an Internet Service Provider, or ISP. ISPs exchange traffic with one another using routers. They also operate email and other servers and provide various means for their customers to connect to those servers. While an ISP may have a multi-gigabit per second pathway on the Internet, it will provide smaller increments of its available bandwidth for the use of its customers. Some popular ISPs are AOL, MSN, and YAHOO!, along with hundreds of others.

The following details various types of ISP-provided Internet connections commonly in use.

Dialup

A modem, often built into a computer, is connected to a standard telephone line. When activated, it dials a programmed number, reaching one of a bank of modems provided by an ISP. The modems convert the digital communications, ones and zeros, into analog sounds, which are carried over the telephone line and converted back to their digital component on the other end.

Once communications between the modems is established, the ISP's router will use DHCP to temporarily assign an IP address to the session.

Dialup is the slowest Internet communication method in common use, with maximum data speeds of approximately 40–45 kilobits per second (kbs). Because of their slow data speed, modem communications are not desirable for video or bandwidth-intensive security applications. Dialup's positive feature is portability, allowing users to connect to the Internet while traveling.

While using a telephone line for modem communications, the line cannot simultaneously be used for regular telephone calls. It's also important to note that in the majority of situations, dialup services are activated for relatively brief periods of time, and then disconnected. During the disconnected time periods, there is no way to remotely access any electronic security equipment connected to the LAN at the client's location.

Definition of broadband

Dialup Internet service is an off-and-on proposition, with a connection being established, communications completed, and the modems disconnected. In contrast, higher-speed "broadband" services are always connected to the Internet, provided that the Internet adapter is physically connected to the network and powered.

DSL

An improved Internet connection based on upgraded telephone lines, called DSL, provides much faster Internet communications than dial-up connections. DSL stands for Digital Subscriber Line, and is actually an offshoot of an earlier technology used for alarm signal communications, which was called "Derived Channel." DSL connections are provided by the installation of a DSL adapter, which connects a local computer, or LAN, to a standard telephone line, which in turn is connected to a Digital Subscriber Loop Access Multiplexer (DSLAM) located in the providing telephone company's central office. DSL service is generally available to customers who live no more than 12,000 feet from the central office, as it may not work properly beyond that distance.

DSL uses a section of the bandwidth available on a standard telephone line to transmit digital ones and zeros, while leaving adequate room for voice communications to be carried simultaneously.

DSL Frequencies

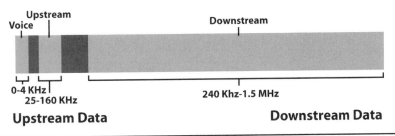

Figure 10-1 Digital DSL data is transported using small segments of the total frequency available on a standard telephone line.

Notice in Figure 10-1 that the transmission bands are separated, so that the voice and data signals do not interfere with each other. There are several different types of DSL service—including symmetrical (SDSL) and asymmetrical (ADSL). ADSL is commonly used for residences and small businesses, while larger businesses may use SDSL. With SDSL service, identical bandwidth is available in both directions. With ADSL there is a much larger bandwidth available for "downstream" data transmission, coming to the subscriber's computer, than is available for "upstream" data going to the Internet. ADSL services were engineered in this manner to best meet typical Internet usage, where a user's single mouse click on a web page (upstream data) produces a large downstream data flow, for example a photo or new web page to be viewed. This can present problems for video applications connected to DSL lines, as video images require large amounts of "upstream" bandwidth to transmit "motion" video over a network.

Standard DSL service typically provides DHCP addressing for the adapter. Upgraded DSL service can provide static IP addresses and increased bandwidth. While both of these upgrades could be useful for general security and video surveillance purposes, they will increase the monthly costs for the DSL service, and may not be available in all areas.

Cable modem   The coaxial cable used to provide multi-channel television for homes and residences can also be used to provide broadband ISP services. A cable modem (which should rightly be called an adapter, as it doesn't modulate or demodulate its signals as a modem does) is connected to the copper coax cable coming into the building, and connected to the local computer or LAN. Just as DSL adapters enable data to share bandwidth with voice communications, cable modems share bandwidth with video (and sometimes telephone) services.

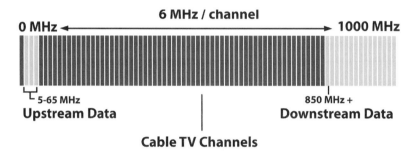

Figure 10-2 Cable modem frequencies. As with DSL, data signals are carried on small segments of the frequencies available in a cable television connection.

As with ADSL, cable ISP service is structured to provide greater downstream bandwidth than upstream. And, like DSL, upgraded services are available that provide increased bandwidth and static IP addresses.

In terms of market share, cable ISP service is leading DSL in the number of US users by a factor of two to one. Recent government statistics (2003) indicate that over 38 million homes and businesses in the US are connected to the Internet via either cable or DSL broadband connections.

> **SECURITY TECHNICIAN'S NOTE:** The effects of asymmetrical Internet bandwidth on security video transmissions must be considered based on how a particular connection is being used. If a DVR, for example, is connected to a DSL connection that has only 128 Kbps upstream, only that relatively small amount of video can sent "up" to be viewed over the Internet. An authorized viewer on the other end of this example may have a large "downstream" capability, but is limited to as much video as can be sent "up" from the transmitting device.

Broadband connections and electronic security

DSL and cable modem connections are very suitable for the implementation of network-enabled CCTV, access control, and alarm transmitters. Because these technologies are usually "always-on," remote access for video viewing or system control can be accomplished without the requirement for any personnel to be on site to activate a connection. These technologies also provide larger bandwidth than dial-up, and remote viewing of reasonable quality video signals can be accomplished.

Alarm clients with broadband connections are perfect candidates for sophisticated alarm and video systems. Low-security digital communicator alarm signal transmission can be upgraded or supplemented with an IP alarm transmitter, which sends its alarm signals over the ISP network in digital format to the central station. These transmitters can be programmed for "polling," where the transmitter and central station regularly exchange communications. Polling assures both parties that the transmitter and central station are prepared for alarm signal transmissions, and can notify the central station and subscriber of line or network failure. More information on IP alarm transmitters is included in Chapter 30.

Network-enabled cameras can be quickly installed and programmed for broadband-equipped clients, particularly if the client has a Wi-Fi router in use. Video servers can connect existing CCTV systems to the Internet using broadband, providing remote access and viewing for authorized off-premise users.

Satellite

In some (primarily rural) areas, neither DSL nor cable ISP services are available. Two-way satellite services, from such providers as DirecTV, can deliver broadband Internet communications to such places, provided that the satellite dish can be installed with a clear view towards the equator.

Satellite Internet service functions like one-way satellite television, at least on the downstream path. A part of the satellite bandwidth is dedicated to providing an encrypted data stream to the subscriber's dish, which carries the data signal to the interface adapter for decrypting and connection to the local computer or LAN. An upstream signal is generated by the interface and passed to the satellite dish, which projects the upstream signal skyward to be received by the satellite. Internet communications are relayed from the satellite to an Earth station, which provides the connection to the terrestrial Internet backbone. Satellite services provide DHCP-type addressing for the client's Earthbound connection. The bandwidth capabilities of consumer-grade satellite connections are from one-half to one-third of the speed of DSL or cable modem services.

The satellite orbits the Earth at a distance of 22, 420 miles from the Earth's equator. This distance causes the satellite to stay in the same position in relation to the earth, even as the Earth rotates through its daily cycle. This is the reason that a satellite dish can be aimed once at the satellite and doesn't have to be repositioned unless it is physically moved or disturbed.

While satellite service is functional for electronic security purposes, it is not preferable if either DSL or cable modem service is an option. Installation of satellite subscriber equipment requires a licensed radio technician, and costs for installation alone can approach $1000 US. Also, this service has less bandwidth and has higher monthly service costs than its competitors. However, it is available just about anywhere, and can provide a broadband connection when no other option can be obtained.

T1 leased lines

Larger commercial enterprises will often opt for a "T1" line, which is a digital offering providing much higher data bandwidth than what is available from DSL, cable, or satellite services. Typically, a fiber optic connection or high-grade four-conductor copper cable is connected from a nearby telco central office to a client's location. At the customer premises is a CSU/DSU (Customer Service Unit/Digital Service Unit) that provides a wired Ethernet output to which the LAN is connected. Full T1 connections provide 1.54 Mbps throughput, generally with high data integrity and reliability. Static IP addresses are available, and many users can be readily connected to one T1 line, providing high-speed data connections for all.

T1 lines would be the preferred choice for the connection of electronic security equipment, provided that it is an available service for a particular client's location. Due to relatively high installation costs and monthly fees, it is most likely that electronic security equipment may share a T1 line with the enterprise computer network.

Virtual private network

Typical data transmissions carried over the Internet or a WAN are not protected; therefore the information is potentially readable by others on the network at the same time. Many organizations use some form of encryption, where data is scrambled using a special mathematical formula called an algorithm. Authorized computers and users can read the encrypted data by employing an encryption key, which is used with a specific encryption algorithm that will restore the scrambled message back to its original form.

Many enterprises want to connect their computer networks together, hooking up networks that may be physically separated by miles. The most secure method to accomplish this is the use of leased lines, but this option is very expensive. Connecting the separate networks over the Internet is attractive from a cost perspective, but can leave sensitive communications vulnerable to interception.

To provide for secure network communications over the Internet or a WAN, the Virtual Private Network, or VPN, was developed. VPNs are often implemented to provide secure communications between remote users and the main LAN of a business, and/or to provide secure data connections between business partners.

A typical VPN consists of a hardware "VPN server" that performs three functions in the connection of outside computers to the enterprise LAN. First, the VPN server provides data encryption and decryption, translating incoming scrambled messages into their original format, and scrambling outbound data packets that will be decrypted by the receiving remote computer. "Authorization" or verification of the identity of remote users is the second VPN server function. Lastly, "Authentication" challenges remote users for their password and username prior to their entry into the network. Remote VPN user computers typically have special client software installed that provides the proper encryption and decryption of data when communicating with the VPN server.

VPN technology can be used over the Internet or on an "intranet," which is a computer network linking multiple buildings using leased lines.

VPN encryption and technology is not standardized; Microsoft and Cisco are the major suppliers to this market, and provide competitive VPN technologies that are not compatible with each other. Network-based VPNs are also available from a number of long-distance carriers.

VPNs and security devices

While it is conceivable that electronic security devices might be successfully connected through VPN devices over the Internet, there are a number of problems with this approach. Most currently produced IP-addressed security cameras, video servers, and alarm interfaces do not have the capability of having the client VPN software installed into them, so they would have to be connected through an existing VPN server to allow communications over the "secure" network. This would then require that the electronic security devices be connected to the enterprise network, which may or may not be allowed by the IT manager. If this type of connection is allowed, the IT manager will have to provide the proper IP addresses and likely will need to manipulate the settings of the VPN server to allow the communications. Another issue is the VPN's authentication process for users and devices. Some electronic security equipment, such as IP alarm signal transmitters, may only transmit data periodically. Such periodic transmissions may be challenged by the VPN server at the receiving end, which may demand a username and password from the transmitter. The IP alarm transmitter may not have the capability within its program to answer such a challenge, causing a failure of communication.

Another problem for security devices connected through a VPN is bandwidth. The process of encrypting and decrypting data packets slows transmission speeds. This would likely have a detrimental effect on video images, which are large files to begin with, and will take more time to encrypt and decrypt than a text file, for example. Some VPN servers have a maximum throughput of only 600 Kbps, or .6 Mbps, which would make the successful transmission of motion video from remote locations impractical. Alarm transmission and access control systems would likely be able to communicate over a VPN, while video signaling most likely will not.

> **WARNING:** If planning the connection of network security devices over a VPN, check with the manufacturers of the security equipment for their recommendations, and consult with the client's IT management to ensure the transmission path and compatibility of all intended devices and connections. Additionally, VPN encryption may cause latency or a delay between the transmission and reception of a data signal, which may cause erratic operations or monitoring of remote IP-addressed security devices.

Internet options

Although there are a variety of Internet connection options, the electronic security contractor will, in most cases, be forced to use the Internet option that has been pre-selected by its client. Clients will be reluctant to increase their monthly ISP/Internet provider costs, unless the contractor makes a case for increasing bandwidth.

Security contractors must also remember that the system they're proposing will be sharing the available Internet bandwidth with the client's business network, which can create problems with IT managers. Remember that the client's business is his main concern, and security may well be a secondary consideration.

Issues of bandwidth requirements and human interfacing with IT personnel are detailed in Chapters 13 and 18.

IP Addressing How-To

One of the basic skills required for technicians to be able to successfully program and install IP network devices is mastery of IP addressing. Every device connected to the network must be properly addressed, or it will not communicate.

Technicians must be capable of checking the IP address of a computer or device, changing the address if needed, and diagnosing address problems. The following section provides instruction about and examples of how IP addressing works and how addressing issues are properly handled.

Addressing overview

Let's review the basics of IP networking:

1. Each device on a LAN must have a unique IP address, different from all others.

2. Each IP address must be in the same network as the others on the same LAN, i.e. the first three address octets must be exactly the same, with a unique number in the fourth position.

3. No number in an IP address octet can be higher than 255.

4. If Wi-Fi and wired Ethernet devices are intermingled on the same network, each device must have a unique address on the network, i.e. a Wi-Fi device cannot have the same IP address as a wired device.

Checking the IP address of a network-connected device

When adding a new component—for example a network camera—to an existing network, the technician may need to find or confirm the IP address of a device already connected. This information is needed so that the new device can be addressed to the same Class C network, enabling communication between the devices.

Now, let's look at some ways to check the IP address of a network-connected device.

There is a set of very useful utilities available from the "Command" line in Microsoft (tm) Windows. Here's how to get there.

Click START, then RUN, then type CMD, and click OK.

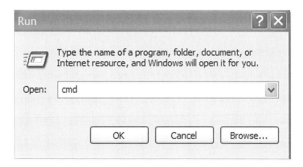

Figure 11-1 Typing "cmd" into the "Run" window, and clicking OK, will open the command line screen.

The command line window will open on your screen. For those familiar with older pre-Windows computer programs, this entry sequence brings the user to a "DOS"-type prompt.

To find the IP address of the computer, type: IPCONFIG and press the "Enter" key. Figure 11-3 provides an example of what you'll see on the screen.

```
C:\WINDOWS\System32\command.com
Microsoft(R) Windows DOS
(C)Copyright Microsoft Corp 1990-2001.

C:\DOCUME~1\DAVIDE~1>ipconfig

Windows IP Configuration
```

Figure 11-2 Typing "IPCONFIG" will show the computer's IP address.

```
C:\WINDOWS\System32\command.com
Microsoft(R) Windows DOS
(C)Copyright Microsoft Corp 1990-2001.

C:\DOCUME~1\DAVIDE~1>ipconfig

Windows IP Configuration

Ethernet adapter Wireless Network Connection 2:

        Connection-specific DNS Suffix  . :
        IP Address. . . . . . . . . . . . : 192.168.1.103
        Subnet Mask . . . . . . . . . . . : 255.255.255.0
        Default Gateway . . . . . . . . . : 192.168.1.1

C:\DOCUME~1\DAVIDE~1>
```

Figure 11-3 The result of typing "IPCONFIG."

What does this command tell us about this particular computer?

The first line of information shows that this computer is using an Ethernet Wireless connection, so it is communicating with the network via Wi-Fi.

The IP address of the listed Wi-Fi adapter is 192.168.1.103. It is often the case that a computer will have more than one network connection, such as both Wi-Fi and wired Ethernet. Each connection type will have a separate IP address. The IPCONFIG command only tells you the address information of the NICs that are presently enabled on that particular computer.

The "Subnet Mask" of 255.255.255.0 indicates that this computer is connected to a Class C network, which is typical of a LAN within a building or campus of buildings.

The "Default Gateway" is the address of the network router, in this case a Wi-Fi router/switch, with which this particular computer is communicating.

Let's try a similar exercise that will provide more information about this computer's IP configuration.

At the command line, type: IPCONFIG/ALL (Notice that there is a space between the "g" and the "/"). Figure 11-4 provides the information displayed.

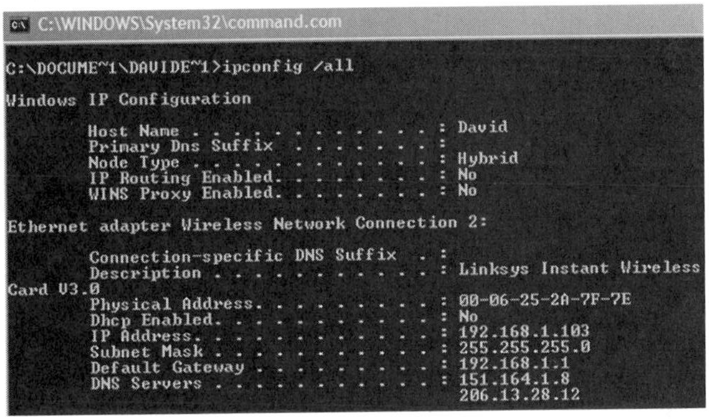

Figure 11-4   Typing "IPCONFIG/ALL" shows the computer's host name, node type, and more.

The additional information provided by IPCONFIG /ALL includes:

"Host Name"—Computers can have a name, which provides an easy nomenclature for locating a specific computer or files on that computer that are accessible. Here the host name is "David."

"Node Type"—As this computer has the capabilities of both wired Ethernet and wireless Wi-Fi, it is a "Hybrid."

Under the "Wireless Network Connection 2" heading, the information provides:

"Description"—Make and model of the Wi-Fi card programmed for that particular connection.

"Physical Address"—This is the MAC address of the Wi-Fi card.

"DHCP Enabled"—This selection indicates whether this particular network interface card will accept IP addresses provided by the system's router/gateway. In this case DHCP is disabled. This means that the IP address for this NIC must be manually assigned.

While the "IP Address," "Subnet Mask," and "Default Gateway" information is the same as with the IPCONFIG command, IPCONFIG /ALL also provides the IP addresses of the DNS (Domain Name Service) servers used by this particular connection to convert alphanumeric names such as www.google.com into their numeric IP addresses.

PING command

The PING command verifies that communications are available between a node on the network or Internet and a particular laptop or desktop.

At the command prompt, type PING (space) and the IP address or name of a network host, such as a PC name on the network or an Internet URL such as www.slaytonsolutions.com. See Figure 11-5 for an example.

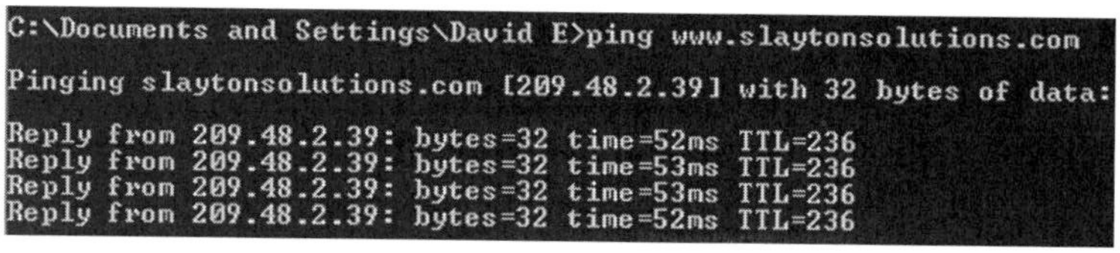

```
C:\Documents and Settings\David E>ping www.slaytonsolutions.com

Pinging slaytonsolutions.com [209.48.2.39] with 32 bytes of data:

Reply from 209.48.2.39: bytes=32 time=52ms TTL=236
Reply from 209.48.2.39: bytes=32 time=53ms TTL=236
Reply from 209.48.2.39: bytes=32 time=53ms TTL=236
Reply from 209.48.2.39: bytes=32 time=52ms TTL=236
```

Figure 11-5   The PING command sends four data packets to the target IP address or URL, which sends responses back for each packet.

The PING command sends four data packets to the target IP address or URL, which sends responses back for each packet. PINGing a device verifies that an active network adapter on the PC can communicate with the other node.

SECURITY TECHNICIAN'S NOTE: Using a variation of the "ping" command, "ping-t" provides a continuous ping function that continues to run until stopped by pressing the "control" key and "c" simultaneously. This is a very valuable test, as it will provide information regarding "latency," or how long it takes a data packet to get from point A to point B in a network or over the Internet. This test also checks for how many packets are lost in communications. While no packets should be lost going over a local network, packets can and do get lost in Internet communications. Any connection that provides more than 5% packet loss is cause for concern.

Exiting command line screen

After completing the inquiry, exit the command line by typing EXIT and pressing the enter key.

Command line functions

As will be shown in Chapter 12, command line functions can provide many quick and easy tools for checking the status of a computer and its network connections.

Finding IP information through Windows

Many computer users find that it is easier to utilize a Graphic User Interface (GUI), pointing and clicking with a mouse, than remembering and typing command line functions.

Windows provides the same information as the command line, and we can access the IP information through the Windows GUI.

To check the IP information from Windows XP, click through the following steps:

START
CONTROL PANEL
NETWORK AND INTERNET CONNECTIONS
NETWORK CONNECTIONS

Figure 11-6   LAN or High-Speed Internet Menu.

In the "Network Connections" window, look under the "LAN or High Speed Internet" heading. Move the mouse over the connection to be checked, and RIGHT CLICK. Move the mouse or cursor down to highlight "Properties," and LEFT CLICK.

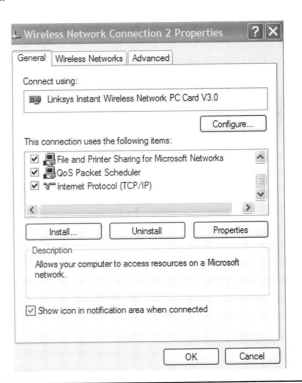

Figure 11-7   Network Properties Screen.

The Network Properties window will open. Use the scroll button on the middle of the right side to pull down the menu of installed protocols until the Internet Protocol (TCP/IP) listing is visible. Highlight this by clicking it, and then click Properties.

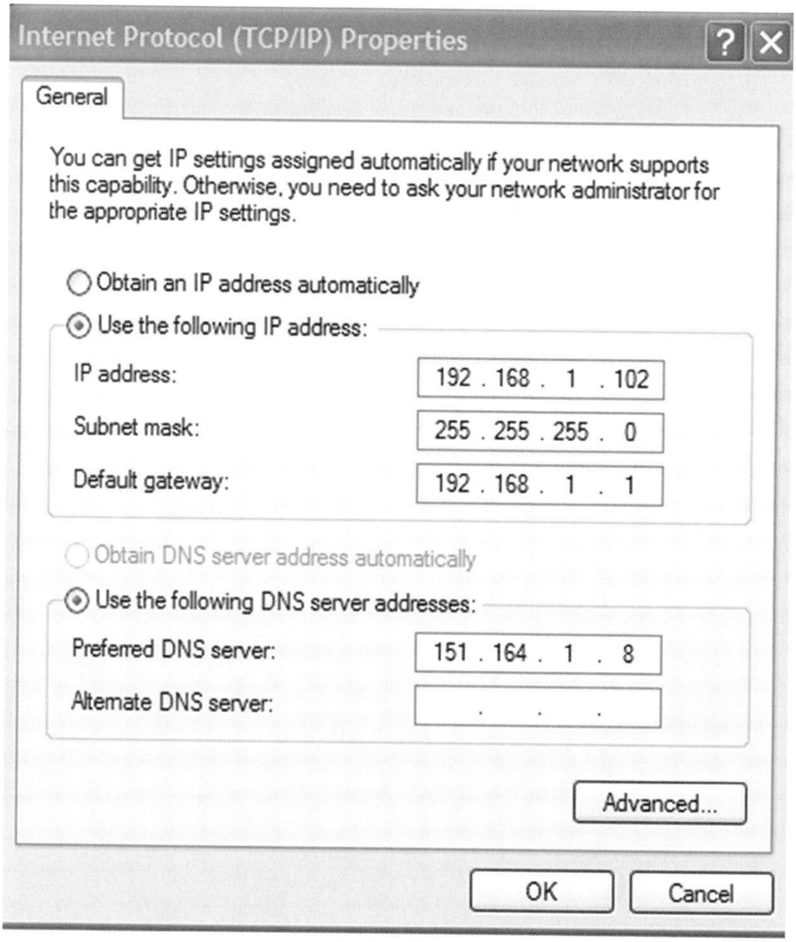

Figure 11-8 TCP/IP Properties Screen.

Here is the same information previously found by using the command line. The IP address, subnet masking, gateway and DNS servers can be changed from this window, if desired. DHCP service can be turned on or off from this screen.

WARNING: If you plan to change the IP address of a computer, or change any of the settings, remember to write down the original settings before performing any changes. You may have to restore the computer to its original connection configuration for it to work properly on the network, and you will need the original settings to do that.

After reviewing and/or changing the computer's network programming, click "OK" to save the changes, or click cancel or the red "X" to exit without saving, and back out of the "Network Properties" windows, clicking "OK" and/or the red "X" in the upper right of the Windows screens.

NOTE: If changing the IP address in Windows XP or Windows 2000, it is not necessary to reboot the computer for it to accept the changed information. Older versions of Windows allow access to and changing of the IP configuration but require the computer to be restarted (reboot) for the changes to be accepted.

Summary

Although specific devices may have different looking screens, the methods of inputting the IP addressing information are similar. The current IP configuration of a PC can be viewed using either the "IPCONFIG" command, or it can be accessed through Windows programs. The "PING" command confirms that one device is accessible by another host on a network. Technicians need to practice changing the IP address in their own computers and in the types of security devices they are commonly using.

CHAPTER 12

IP Addressing Example

To demonstrate the complexities of IP addressing, the following details the different IP addressing schemes used in a typical small network, such as that used at the home office of the Security Networking Institute. As will be detailed in this example, both dynamic (DHCP) and static IP addressing can be used within the same network.

Equipment Layout

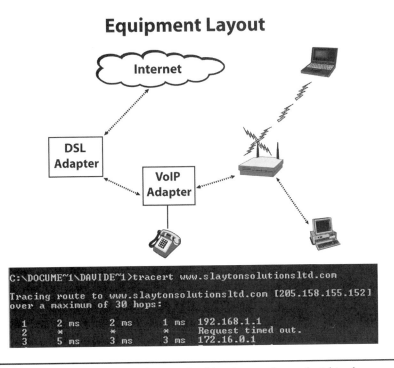

```
C:\DOCUME~1\DAVIDE~1>tracert www.slaytonsolutionsltd.com

Tracing route to www.slaytonsolutionsltd.com [205.158.155.152]
over a maximum of 30 hops:

  1    2 ms     2 ms     1 ms   192.168.1.1
  2    *        *        *      Request timed out.
  3    5 ms     3 ms     3 ms   172.16.0.1
```

Figure 12-1   Both dynamic and static IP addressing can be used within the same network.

As seen in the above illustration, this network consists of a DSL adapter providing connection to the Internet, a VoIP adapter that provides telephone service, and a Wi-Fi gateway router that provides wired and wireless network connections for a PC and a laptop computer.

DSL IP addresses

As a dynamically-addressed DSL adapter, this device receives its WAN or "public" IP address from the ISP's server, in this case SBC-Yahoo.

Using the **www.whatismyip.com** Internet program, we can see that the current WAN IP of the DSL adapter is 68.72.101.129. As detailed in Chapter 10, this address is dynamic and can change at any time.

For local communication, the DSL adapter can be accessed from a directly connected PC, which is set to accept a DHCP address from the DSL adapter. The IP address 172.16.0.1 is used as the DSL adapter's local area network IP address. This private LAN address is static, and is the default local IP address of this particular device.

The private IP of the DSL adapter may not be readily apparent, as the adapter is often accessed by using an icon or desktop shortcut on a PC on the network. Using the TRACERT (trace route) command line function, the LAN address of the DSL or cable modem can be determined.

In Figure 12-2, TRACERT is used to discover the routers between a PC and www.slaytonsolutionsltd.com.

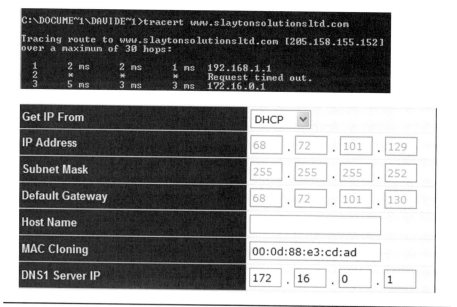

Figure 12-2 TRACERT is used to discover the routers between a PC and a URL.

The first IP address listed is the gateway router's LAN address, that being 192.168.1.1. The second IP address visible is the DSL adapter's LAN address, in this case 172.16.0.1.

VoIP IP addresses

Moving on to the next device connected, Figure 12-3 shows the WAN addressing screen from the VoIP adapter.

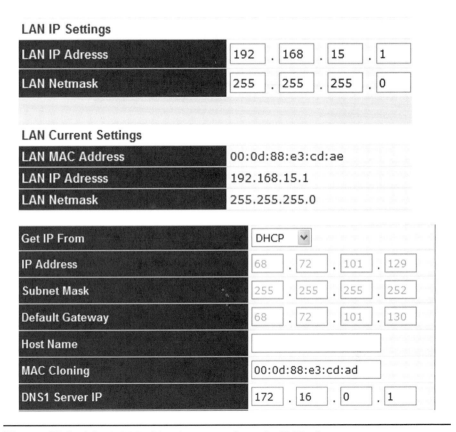

Figure 12-3 The WAN addressing screen from the VoIP adapter.

Notice the WAN address is set by DHCP, which is being fed by the DSL adapter. Also notice that the WAN address is the same as the Internet IP address of the DSL adapter, that being 68.72.101.129.

The reason the WAN addresses are the same for both the DSL and VoIP adapters is that the VoIP adapter must be placed in the "DMZ" (Demilitarized Zone, see Chapter 22) of the DSL adapter's firewall to function properly. This setting connects the "DMZ'ed" device directly to the Internet, and allows any and all traffic to and from the VoIP adapter. This is the reason that the WAN addresses are the same for both the DSL and the VoIP adapters.

Now let's look at the LAN addressing of the VoIP adapter.

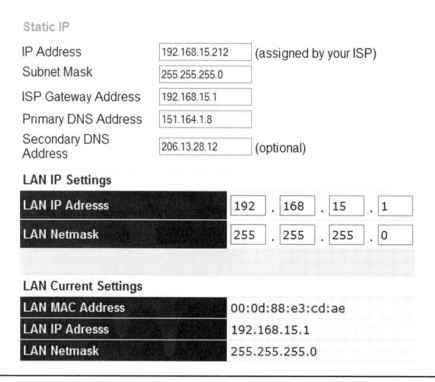

Figure 12-4 The LAN addressing of the VoIP adapter.

Figure 12-4 shows the LAN addressing of this device to be 192.168.15.1. This again is a static default address, and this device offers the option of changing this address if desired.

> **SECURITY TECHNICIAN'S NOTE:** The Internet Assigned Numbers Authority (IANA) has assigned the following IP address ranges for "private" network use:
>
> Class A: 10.0.0.0-10.255.255.255   Subnet: 255.0.0.0
> Class B: 172.16.0.0-172.31.255.255   Subnet: 255.255.0.0
> Class C: 192.168.0.0-192.168.255.255   Subnet: 255.255.255.0
>
> While many IT departments use the 192.168.0.0 range for their local networks, it is not unusual for an enterprise to use either of the other address ranges for their local devices.

Wi-Fi gateway router IP addressing

Turning to the next connected device, Figure 12-5 shows the current IP addresses of the Wi-Fi router.

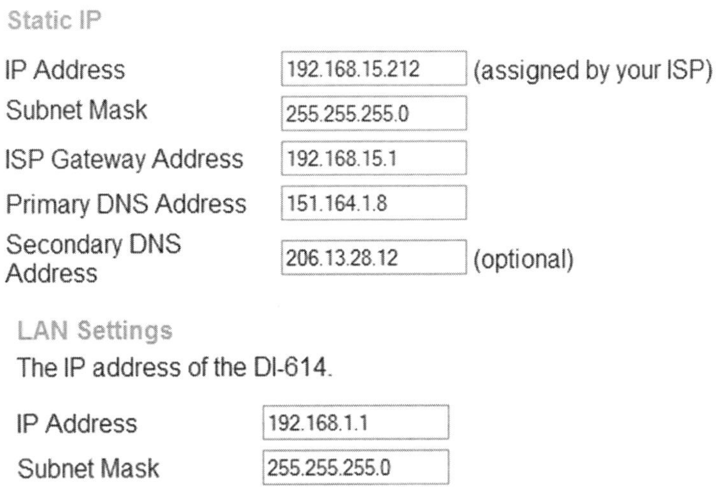

Static IP

IP Address	192.168.15.212	(assigned by your ISP)
Subnet Mask	255.255.255.0	
ISP Gateway Address	192.168.15.1	
Primary DNS Address	151.164.1.8	
Secondary DNS Address	206.13.28.12	(optional)

LAN Settings

The IP address of the DI-614.

IP Address	192.168.1.1
Subnet Mask	255.255.255.0

Figure 12-5   The current IP addresses of the Wi-Fi router.

Here the Wi-Fi router has been programmed for a static WAN IP address, 192.168.15.212. Notice that the "ISP Gateway" address for this particular router is 192.168.15.1, which is the LAN address of the upstream VoIP adapter.

Figure 12-6 shows the LAN settings for the Wi-Fi router.

```
C:\WINDOWS\System32\command.com

Microsoft<R> Windows DOS
<C>Copyright Microsoft Corp 1990-2001.

C:\DOCUME~1\DAVIDE~1>ipconfig

Windows IP Configuration

Ethernet adapter Wireless Network Connection 2:

        Connection-specific DNS Suffix  . :
        IP Address. . . . . . . . . . . . : 192.168.1.103
        Subnet Mask . . . . . . . . . . . : 255.255.255.0
        Default Gateway . . . . . . . . . : 192.168.1.1

C:\DOCUME~1\DAVIDE~1>
```

LAN Settings

The IP address of the DI-614.

IP Address	192.168.1.1
Subnet Mask	255.255.255.0

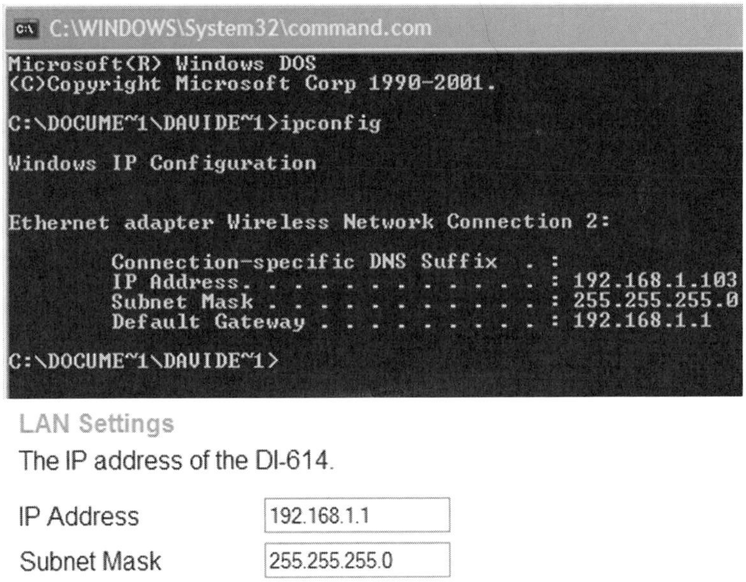

Figure 12-6   The LAN settings for the Wi-Fi router.

The local IP address is static and set for 192.168.1.1. This address then becomes the "default gateway" address in the connected PC and laptop computer.

Computer IP settings

Figure 12-7 shows the IP address screen for the laptop computer that communicates with the network through the Wi-Fi router.

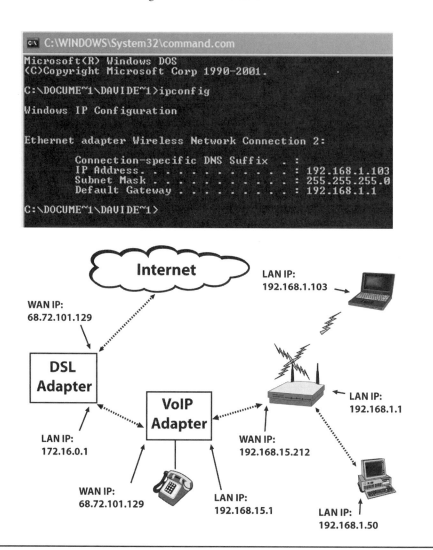

Figure 12-7 The IP address screen for the laptop that communicates with the network through the Wi-Fi router.

The laptop's IP address is 192.168.1.103, and its "default gateway" address is the Wi-Fi router's LAN address.

Putting it all together

Figure 12-8 shows the original equipment drawing, with the WAN and LAN IP addresses included.

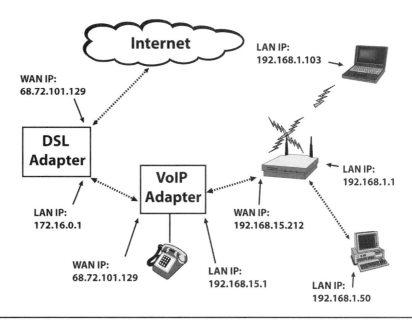

Figure 12-8   Both dynamic and static IP addressing can be used within the same network with WAN and LAN addressing.

So even in a very small network, there can be a number of addressing schemes. In this case, the DSL adapter's Internet IP address is 68.72.101.129. The DSL adapter passes the "outside" Internet IP address to the VoIP adapter, which in turn uses a separate sub-network, IP address 192.168.15.1, to communicate with the connected Wi-Fi router. In turn, the Wi-Fi router uses a separate sub-network address, 192.168.1.1, to communicate with the desktop and laptop computers.

Gateway routers and LAN/WAN addressing

As can be gleaned from this exercise, the addressing of IP devices is very exact, and follows certain rules. One of the defining features that designates a gateway router is having two IP and MAC addresses, one for the LAN and one for the WAN. In this very small network, there are actually three gateway routers; the DSL adapter, the VoIP adapter, and the Wi-Fi router. Each of these devices talks to its upstream and downstream connections using a separate sub-networking address. So the DSL adapter talks to the Internet using 68.72.101.129, while passing that same address connection to the VoIP adapter, which sits in the DMZ. The VoIP adapter communicates with the Wi-Fi router using the 192.168.15.1 sub-network address, and the Wi-Fi router talks to the connected computers using 192.168.1.1.

As these sub-networking address schemes can be difficult to fathom, the benefits of DHCP usage become readily apparent. When installing this network, the owner or technician can labor over plotting and planning the correct sub-network addresses, or simply enable DHCP in every device and let the machines figure it out themselves.

Summary

Dynamic IP addressing (DHCP) is commonly enabled on network equipment, providing automatic addressing and configuration as devices are connected onto a network. Both DHCP and static IP addresses can reside on the same network at the same time. As "adapters" such as DSL and VoIP are connected, they will set up their own separate LAN network addressing to communicate with adjacent devices. Understanding these separate network addresses is critical for security technicians. To reach a specific device on a network, the IP address must be known.

CHAPTER 13

Working with the IT Department

Many applications of network-enabled electronic security systems will occur at commercial or "enterprise" locations. As one of the key benefits to networked security systems is the ability to use the existing network infrastructure and equipment, an understanding of the IT (Information Technology) manager's desires and motives is critical to the alarm dealer's success in smoothly integrating the video and/or security transmissions into the greater mix of enterprise communications. Understanding the IT person's perspective allows electronic security dealers to best craft their discussions and presentations of new technologies, and increases the potential cooperation of client personnel, who can make or break the sale of the system and the installation of same.

If we conceptualize the existing network as an interstate highway, the client's IT management function as traffic cops, allowing or disallowing traffic on "their" piece of road. Understanding their requirements, and being able to explain what impact the new equipment may or may not have on network traffic, will go a long way toward smoothing out what can be a contentious relationship between the alarm company and the IT department.

IT management responsibilities

To best understand the IT person's concerns, it's necessary to understand what his or her responsibilities consist of. The IT management job is not just to manage the cable plant, computer hardware, and software used in the enterprise network; they must also manage their subordinates who work on the system. Another overriding responsibility of IT management is to function as the human interface for all of the users of the computer system, which usually means nearly every person in the company. If and when users have problems with their computers, they will be calling the IT department for assistance.

IT is also responsible for managing a budget for the purchasing of hardware, software, and computer-related services. This is a "hidden" reason for getting the IT person on your side; they often have the largest budget in the company. If they can be sold on the benefits of a networked security solution, monies from the IT budget may be accessed. It is conceivable that plans for a new network security system may require new switches, routers, or other equipment that the IT person has been wanting to acquire, making the implementation of the security system attractive to them.

IT management concerns

Constant pressure is placed on IT management to provide, install, or allow new software and hardware products for system users. Many individuals are computer-savvy, and may see a new product at a show or in a magazine and say "We need that!" for the computer network that they use at the office. From the IT manager's perspective, any new product, software, or service that is placed on his or her network must be supported. They have to understand it, be able to troubleshoot it, and explain its functions to network users. With reduced personnel counts in their department and the uncertainty of functionality, IT people tend to be reluctant to accept new devices on their network. Their easiest answer is "No," and it may require pressure from upper management to change their opinion.

IT management are powerful people in a business or organization. They have the power to allow or disallow equipment, products, and services onto their network. Although a brand-new separate and private cabling and hardware network can be installed to provide connectivity for a networked electronic security system, it will be much faster and cheaper to use existing network paths for security communications. And use of existing network paths will involve IT personnel, who must be carefully cultivated by the security dealer.

There is no positive outcome that can result from a rancorous or adversarial relationship between the electronic security company and IT management. If a networked security system is installed without IT's blessing, it will be a simple matter for IT personnel to make the security company's servicing task a nightmare after the installation is complete. Remember, they're at the job site constantly, with the ready knowledge and ability to cause problems with the security system. A few unplugged cables, reprogramming of a switch or router, or closing of software ports, and the security system is "off the air." Who's going to get the call to fix the problem? You are.

Successful implementation of a network security system integrated onto the customer's enterprise network requires the active and positive involvement of the enterprise's IT department.

IT and network security

One of the most important responsibilities of the IT department is the protection of the enterprise's network and data from hackers, both outside the organization and within. This computer network security issue is another reason why IT people may be reluctant to embrace the installation of IP-addressed cameras, video servers, and alarm transmitters onto their system. The perception of the IT department may well be that new devices provide new security "holes" that must be closed, managed, or monitored.

The successful electronic security company will express its concerns for maintaining the security of the enterprise network, while installing new equipment that will provide increased physical security for the client's property and personnel. Being able to "talk the talk" about network and data security with the IT manager will help convince him or her that the proposed security system can help to increase the level of security, not diminish it.

How to work with the IT department

Once upper management has been presented the concepts of networking electronic security, for example connecting a campus of buildings into a central monitoring station, it's important to involve IT management very early in the sales cycle. For all of the reasons listed above, the "buy-in" by IT of the proposed system is critical to its success. If the relationship is positive, there may well be unexpected benefits that can produce a networked security system that is lower cost, higher security, or both. For example, the IT department may be aware of fiber or copper cables that were installed after the original structured cabling plant was put in place. If there are unused conductors in those cables, these may be quite suitable for networked security communications, potentially saving thousands of dollars in installation costs.

During these initial discussions with the IT personnel, it is very important that the security company's representative have a thorough knowledge of bandwidth limitations and how they can affect the performance of the enterprise network. IT departments constantly monitor the bandwidth capabilities of their network, and are often presented with complaints from network users about slow communication speeds. We live in a world where what was an acceptable speed for computer networks two years ago is now too slow. Bandwidth management is a juggling act that IT personnel perform constantly in their attempt to provide all network users with a reasonably fast network.

Bandwidth issues come to the forefront when it is proposed that security video signals be transmitted over a network. "Video" usually means large bandwidth requirements, and IT personnel may be fearful that the video streams may slow down enterprise computer functions.

Going parallel

The easiest and best way to provide a top-notch networked security system is to use a "parallel" network cabling system, as described in the next chapter. By utilizing some otherwise unused conductivity, the security signals can travel within a building or around a campus without taking any bandwidth away from the enterprise network.

Using the enterprise network

If the preferred parallel network concept is unavailable, the security devices must be connected to the enterprise network to allow signal transmission and access to security equipment. Connection of access control, alarm signal transmitters, and similar devices will have little or no impact on the performance of the enterprise network, as these devices require little bandwidth, and more importantly, will only be using the network periodically.

Network video signals are a different matter, as they can require substantial bandwidth and will generally be transmitting their images over the network on a constant basis. In this situation the electronic security salesperson or project manager needs a complete understanding of what bandwidth controls are available on the products or system they are proposing, and how to balance the need for quality security video images with the availability of network bandwidth.

Bandwidth controls

As is detailed in Chapters 17 and 18, the quality of the image, number of frames per second, and compression algorithm used will determine the approximate bandwidth requirement of a security video feed. Remember that video bandwidth requirements may fluctuate based on the type of image and amount of movement within the camera's viewing field.

It's important to know that when connecting to a wired Ethernet network, the concentration of multiple cameras or video servers onto backbone segments can add up to heavy bandwidth requirements. For example, if five video servers, each using 2 Mbps of bandwidth, are connected to a network switch in a horizontal cross-connect, 10 Mbps of bandwidth in the backbone will be needed to transmit those video signals to the main cross-connection point. If the backbone operates at 100 Mbps, now 10% of the backbone's capability is being used for security video feeds. However, if the five video servers were each connected to different horizontal cross-connects, each would only be taking up 2% of its connected backbone's capacity. For networks that operate at 1000 Mbps for backbone connections, the addition of a number of 2 Mbps security video server feeds should be of minor concern to IT management.

If the enterprise is the 9–5, closed-on-the-weekends type, it's possible that the video servers, DVRs, or IP cameras may be programmed to transmit fewer frames per second during the workday, and go to maximum frame rates during off-hours. This can be done manually by connecting to each remote IP security device when bandwidth changes are needed. As networked security devices become user-friendlier, this type of variable speed transmission may well become a pre-programmable scheduled function.

> **SECURITY TECHNICIAN'S NOTE:** When dealing with an IT manager who's reluctant to allow security video to be placed on their enterprise network, an effective way to sway his or her opinion is to provide a network video camera (or camera + encoder) of the type proposed for the installation for the IT manager's experimentation. Bring the IT person a network camera, lens, power supply, and any necessary software, and help him or her set it up on his or her network as a test. Sophisticated IT departments have software that can monitor the bandwidth usage on their network; the test camera/encoder will let them see what bandwidth will be used by the video devices.

How much is available?

Another approach to the issue of video bandwidth over the enterprise can be to ask IT management how much bandwidth can be allocated for video purposes across the network. This concept can help involve the IT personnel in the early stages, as the security contractor can work with IT to plan out the best implementation of the networked video system. The benefit of this approach is that the IT personnel can develop some "ownership" of the project, as their bandwidth design is included. After determining the bandwidth capabilities, the security contractor can either determine what equipment meets the requirements, and/or how to "throttle" IP cameras and video servers so as not to exceed the planned bandwidth usage.

What about the future?

If a positive relationship has been developed with IT personnel, the astute security system designer will be thinking and talking about the potential for future expansion of the system and what impact it may have on the enterprise network. One of the benefits of having a good rapport with IT management is that they may well help in the sales of additional security system components to the client by providing usage of additional network paths and bandwidth. If working closely with a client's IT management resulted in a smooth running system for their headquarters, for example, it may then become a simple sale to connect systems from their outlying buildings, provided that the IT manager is on board.

Summary

If the network administrator is on your side, network security system sales and installation can be as easy as possible. If the IT department isn't on your side, any sizeable network system may be in deep trouble.

Here are the points to remember when working with the IT department:

1. If possible, use a parallel cable network to connect security devices, as detailed in Chapter 14. This eliminates any bandwidth usage conflicts. Always check for this option, particularly for network video systems.

2. Involve IT personnel as early as possible in the sales process. Make the network security project "their" system, and use their input to best configure devices.

3. Understand bandwidth issues, and be able to discuss them intelligently.

4. Be knowledgeable about various product lines and their bandwidth usage or requirements and controls or adjustments.

Cabling and Connection Options

When planning the connection paths of a network security system, installing companies can be confronted with a variety of potential cabling options. One of the great strengths of networked security systems is the ability to connect devices to existing cabling, reducing installation time and costs. How to connect to existing networks, and the potential pluses and minuses of different configurations, will be explored in this chapter.

Standardized structured cabling

As was previously detailed, most structured cabling installations that a security firm might encounter will generally conform to the standards of EIA/TIA 568. These standards dictate how many wall outlets are to be installed, the amount of conductors to be present in the telecommunications closets, and so forth. In the following examples, it is assumed that the EIA/TIA standards have been followed by the contractor who installed the cabling. However, it's important to remember that unless careful testing and system certification was performed, it's quite possible that parts of the cabling system may be defective, not connected, or have some other problem that may cause a portion of the cabling network to be non-functional. *On this topic, it is important for the security contractor to spell out in the client's contract exactly who is responsible for repairs in the event that pre-installed cabling that is to be used for the security system is defective.*

Cabling as it stands

Figure 14-1 shows the typical structured cabling plant, bringing computer and telecom connectivity to locations throughout a building.

The "main cross-connect" is where the Internet connections and telephone lines enter the building, and is usually located in the basement or first floor of a building.

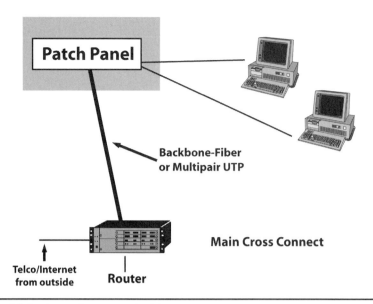

Figure 14-1 Typical structured cabling.

This room may also contain network computer servers, UPS power and other devices and services necessary for the functioning of the enterprise's computer network. As such, the main cross-connect room may be an excellent choice for the location of security-specific powered devices, due to the potential availability of power backup and proximity to Internet connections.

> **SECURITY TECHNICIAN'S NOTE:** Be sure to get authorization from the customer before connecting to their backup power system. There needs to be adequate capacity in the emergency power system to add the devices you wish to connect.

Backbone cabling

The next key element in a structured cabling system is the "backbone," which is a cable connecting a remote "horizontal" cross-connection point to the main cross-connect. Backbone cables may consist of multi-pair UTP copper conductors, which may be Category 3, 5, 5e, or 6 rated conductors. While Cat 3 cable is adequate for voice communications and 10 Mbps Ethernet, it is not recommended for 100 or 1000 Mbps. The electronic security contractor should avoid connection of remote network devices, such as cameras or video servers, to Cat 3 cables. Cabling rated higher than Cat 3 should be adequate for security system use, provided that the cables were properly installed, terminated, and tested.

Fiber optic cable is often used for backbone connections, with the glass fiber being connected to "media converters" that retransmit electrical Ethernet data as a blinking light, which passes through the fiber and is reconverted at the other end. Fiber is used for backbones as it provides great distances of communication, immunity to radio frequency interference (RFI) and electromagnetic interference (EMI), and tremendous bandwidth potential. Generally, fiber cables installed between connection points in a structured cabling system have a number of unused spare fibers available for alarm contractor use.

What's important to know is that accepted cabling standards recommend that spare conductivity be provided between the main cross-connect and the horizontals, to enable future expansion of the network. So there may well be currently unused conductors between these locations, either fiber or copper pairs, which can allow the security contractor to connect security devices in a manner that isolates security video and communications from the main enterprise network.

There may well be multiple horizontal cross-connection points, all cabled back to the main cross-connect. These multiple locations may be on every floor or every other floor in a multi-story building, or may be situated in different buildings in a campus environment.

Horizontal cross-connect

The intermediate or "horizontal" cross-connection is a terminating center, where cabling from work areas is connected to the overall network. Usually located in a "telecommunications closet," a horizontal cross-connect consists of patch panels, media converters, and other network equipment. These centralized connection points allow network administrators to quickly change connections and services when new users need network and telecom connectivity. As there are usually powered enterprise network devices housed in the horizontal cross-connection room, there is often a UPS power supply. This is another potential installation location for a security contractor's network equipment.

Horizontal cabling

Leaving the horizontal cross-connect room, cabling is routed around the users' areas, providing computer and telephone connections for work stations. In almost all cases the cable will be four-pair copper UTP, with a minimum of two such cables per work area. These copper cables will be terminated onto female RJ-45 eight-wire ports that can be wall mounted or, in some cases, included in partitions or office desk furniture.

In rare cases, fiber optic cable may be extended out from the horizontal cross-connection to provide direct fiber connections for high-bandwidth users and applications. According to the EIA/TIA, horizontal cabling is not to exceed 100 meters (328 feet), including any short "patch cables" that connect network devices or telephones onto the network.

Network camera installation example

Now that the typical cabling system has been reviewed, a typical security system installation can be planned. In this case, the cabling options for installing a network camera will be detailed. This camera will be installed in the work area and viewed on a computer in the main cross-connect, with optional viewing over the Internet.

Although the example in Figure 14-2 uses a network camera, the same principles apply to any type of IP-addressed electronic security device, such as video servers, access control interfaces, or alarm transmitters.

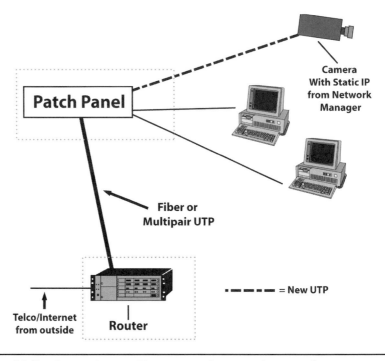

Figure 14-2   Connecting new IP camera to the enterprise network.

Although there may be unused RJ-45 sockets available in the work area, and the temptation great to connect the camera to them, this should be avoided. It is just too easy for a person to unplug the camera's network connection from the wall, and it's also probable that at some point the client will need that particular network socket for the enterprise network or a telephone connection. The security contractor should install new UTP cable from the camera location to the horizontal cross-connection room, placing male RJ-45 connectors on each end to plug into the camera and the patch panel within the horizontal cross-connect.

Camera power   Many of today's network cameras are powered by low-voltage AC, supplied by a plug-in transformer. Security installation technicians need to carefully plan for such power connections, making sure that the AC outlet selected is not on a switched circuit, which can kill power to the camera. *Care should also be taken in selecting the location for that plug-in transformer. It should not be easily accessible, which would allow someone to inadvertently or deliberately unplug the transformer.* Some cameras can be powered with low-voltage DC current, which in some cases can be transported on the unused pairs of the UTP cable connecting the camera to the cross-connection location. Security dealers should understand and carefully follow the manufacturer's instructions if using an "up the cable" powering option.

> **SECURITY TECHNICIAN'S NOTE:** As explained in Chapter 20, Power over Ethernet ("PoE") is a standards-based technology that can provide DC power for cameras, encoders, and other devices on the same UTP cable that connects the device to the Ethernet network. Many vendors are presently supplying their cameras with PoE capability. It's likely that most network-enabled security devices will have PoE capability in the near future.

Connection to the network

If given the OK by the client's network administrator, the network video camera can be connected to an unused patch panel port in the horizontal cross-connection room. Assuming the selected patch panel port is properly terminated, the network camera is now functional on the network, provided it has been programmed with a network-compatible IP address. In the case where the camera is being connected to the network along with other enterprise computers, the system administrator will provide a correct IP address for the security dealer to program into the camera.

Wait now

As will be detailed later, network administrators may be loathe to allow the connection of "foreign" devices on their network, as they are concerned about the security of their network, and bandwidth usage. If the electronic security company is denied access to the common enterprise network for its devices, the use of "parallel" network cabling can provide the security dealer with the benefits of using some of the existing cabling in the client's location, while increasing the security of the alarm/surveillance/access control systems' communications.

Parallel networking

To conceptualize parallel networking, consider an interstate highway connecting two interchanges. While there is the main highway, perhaps three lanes in each direction, there is also the little-used two-lane "access road" running along both sides of the highway, allowing local traffic and an emergency route in case of problems with the road or an accident.

The same concept is available in structured cabling systems that were installed with currently unused conductors, either fiber or copper, between the horizontal cross-connection points and the main. The "access road" between the connection points may well be unused, readily available, and perfect for security system network connection. And contrary to the highway example above, the unused conductors between connection points will likely be of equal potential bandwidth to the main network.

Benefits of being alone

Isolation of the electronic security network devices from the client's enterprise network provides many benefits. As the signals are isolated, there is no impact on the enterprise's bandwidth usage . . . the security system can't be blamed for network slowdowns. An isolated security network is more difficult for insiders to hack, as their computers are not electrically connected to it, so inside personnel cannot easily dial up a security device to see how it works. Politically, an isolated security network can produce better relations with the client's IT manager, as he or she is not forced to take responsibility for the connectivity of unfamiliar equipment. The network manager provides the use of otherwise unused conductors between connection points, and the security installation company takes responsibility for all addressing, networking issues, etc. for the security devices. In many ways parallel networking of security devices is preferable to sharing bandwidth with the enterprise. When planning and installing a parallel network, the security contractor must provide, install, and possibly program any needed hubs, switches, routers, or Internet connection adapters.

Going parallel-testing

The first step is to find out if unused backbone conductivity is available, and to obtain the approval of the client's network manager for its use. The next step is to test the quality of the proposed conductors, whether copper or fiber. In the case of UTP, proper terminations and bandwidth capability can be verified with a fairly inexpensive tester ($125 or so), that will confirm that the jack sockets at each end have been properly connected, and that the cable can pass 10/100 Mbps Ethernet. This assumes that the spare conductors have been previously terminated onto RJ-45 jacks at both ends; if not, female jack sockets will need to be installed.

If the conductors offered are fiber, shining a standard flashlight into one end of the fiber can confirm that the fiber isn't internally broken, provided that the light comes out of the other end of the fiber being tested. If the light doesn't pass through, it is likely that one or both of the fiber connectors were poorly installed and/or defective, and needs replacement.

> **WARNING:** It is common for currently unused fiber strands to be either unterminated or to have poor quality connectors installed by the original contractor. Remember that these fiber strands were planned to be spares, and they likely have never been connected to the network, so their potential defects have never been exposed.

Parallel network cameras example—copper

Once it has been determined that parallel conductors are available between the horizontal cross-connect and the main cross-connect, and the conductors have been tested for suitability, Figure 14-3 provides an example of the installation of network cameras on a parallel network, connected to available UTP backbone pairs.

The horizontal UTP is pulled from the camera location to the telecommunications closet, with male RJ-45 plugs installed on each end. If only one camera is being connected to that particular horizontal cross-connect, its cable can be plugged directly into the parallel backbone UTP for data transport to the

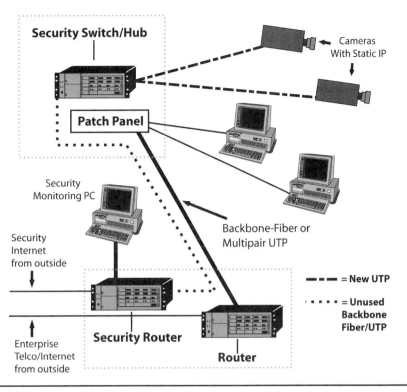

Figure 14-3 Parallel network cameras.

main room. If more than one camera, video server, or IP-addressed security device is being connected in one closet, the installing company would place a hub or switch with sufficient port quantity in the closet. This device requires AC power, and its transformer would preferably be connected to a UPS.

NOTE: Switches are preferable to hubs, as they provide larger bandwidth potential for signal transmissions. Small five-port hubs and switches are very inexpensive, less than $50 US, and are perfectly suitable for these types of applications.

Once the cameras are connected, properly addressed, and programmed, their video signals will be available on the monitoring computer, once it is connected to the selected backbone UTP within the main connection room.

NOTE: This example is provided to explain the concepts of parallel networking, and some aspects of the installation have been simplified. For example, most network cameras function as web servers, and their images can be viewed from any standard web browser software, such as Netscape or Internet Explorer. However, often system or camera-specific software must be installed in the viewing computer to provide the video decompression algorithm program necessary to view a particular brand of network camera. These issues will be detailed in a later section of this guide.

Fiber hookup

The connection to available fiber backbone conductors is similar to the UTP example above, with the addition of a pair of fiber-to-UTP "media converters," which are simple non-programmable devices that retransmit Ethernet UTP signals onto fiber, and change the optical signals in the fiber to electrical Ethernet data on the other end. Fiber media converters are shown in Figure 14-4.

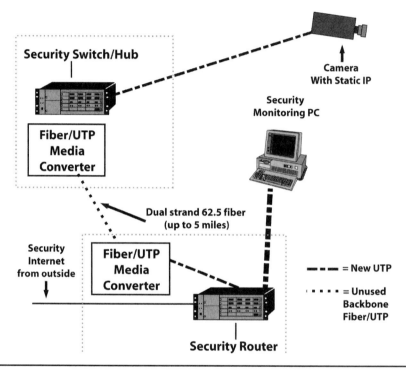

Figure 14-4 Media converters.

As with the UTP parallel system, if more than one security device needs to be connected within a single horizontal cross-connection room, the security company will need to provide a switch or hub with the proper number of ports. Once the cameras are connected via UTP to the switch/hub, a UTP patch cable is used to connect the switch to the media converter.

Fiber media converters

There are many manufacturers of fiber-to-UTP media converters, and each provides many different devices. The proper selection of a media converter requires knowledge of the type of fiber in place, how many fibers the client will allow you to use, and the overall length(s) of the fiber cables to be used.

"Multimode," fiber is used for distances less than four miles. Multimode fiber will be labeled on the jacket as "62.5/125" or "50/125," which indicates the sizes of the fiber core and cladding in microns, which are millionths of a meter.

Fiber labeled "9/125," "8/125," or "10/125" is the "singlemode" type, which is used for long distance and high-bandwidth applications.

Media converter models are differentiated by the type of fiber (multimode or singlemode), number of fibers needed (one or two), the maximum distance over which they can communicate, and whether the devices will transmit 10, 10/100, or 1000 Mbps Ethernet. Fiber media converters that use two fibers are less expensive than those that use a single fiber for communications.

Once a media converter model is selected, remember to order one for each connection end, along with appropriate power supplies.

Viewing options

Once the network cameras, video servers, or other electronic security devices are connected and communicating on the LAN within the building, various options are available to provide viewing and control capability for authorized users. It is a simple matter to connect a single computer to the parallel network to be used as the primary viewing and control station. This would typically be located in the guard station. If additional computers within the building are to have access to the security devices, they can be connected by installing or connecting separate UTP cable, which is then connected to the parallel "security" network via an available hub or switch port. Each computer also will need an additional separate NIC card. Authorized personnel could then use any of these computers to connect to the enterprise network or, by selecting the "security" network from the computer, could also view and control the electronic security system. There may be limits on the number of clients that can connect simultaneously to a camera or video server.

Pre-installation testing of parallel systems

One of the great advantages to parallel networking of electronic security systems is that the entire system can be connected, programmed, and tested at the dealer's office before being installed at the client's location. Video servers, network cameras, and other IP security devices can be addressed, programmed, and connected to a temporary network made from simple UTP patch cords. Such pre-testing will expose any problems with the programming and/or connections, and these problems can be resolved before the installation starts, lessening installation time and problems at the job site. If the system has been pre-tested, network cameras and other IP security devices can be labeled and numbered for rapid and sure installation into the client's location.

10/100 problems

The Security Networking Institute has tested many network security devices, along with commonly available hubs, routers, switches, and fiber media converters. During these tests there have been occasional problems with specific devices not passing their data through network hardware such as switches. The typical problem found is that some 10 Mbps devices will not transmit their signals through some hubs and switches, even those that are labeled as "10/100 Ethernet compatible." This can cause problems for electronic security installations, particularly those involving the connection of security devices to enterprise network hardware provided by the client or already in place. Many of the high performance enterprise switches either will not pass 10 Mbps Ethernet data, or may need to be reprogrammed to provide this function. This problem also appears when using inexpensive fiber media converters, as the least expensive units only transmit 100 Mbps, while many security devices only transmit at 10 Mbps.

These potential problems emphasize the importance of pre-programming and testing all electronic security devices, switches, and monitoring computers before taking them to the job site. In the event that the security dealer is planning to connect security devices to the client's switches, hubs, or other network hardware, the planned devices can be pre-programmed with a correct IP address and temporarily plugged into the network. If the device can be "pinged" from another computer on the LAN, this indicates that it can be reached.

Security of telecom rooms

As security equipment will be installed into remote horizontal cross-connection rooms, it is important that such rooms be secured from unwanted access by potential intruders. Doors should be securely locked, access to the room should be electronically controlled, and the door should be connected to an intrusion alarm system that monitors unauthorized opening. These communications rooms should also have fire detection devices in place that are connected to a fire alarm panel. Security dealers should see this as an opportunity to sell additional products and services to the client; door contacts, access control, and other alarm communications can be connected to inputs on some video/alarm servers, transmitting alarms from the communication room to the monitoring station over the IP network.

Summary

Standardized structured cabling systems provide a wealth of options for the connections of electronic security devices. Such components can be connected directly to the enterprise data network, or there may be the option of using a "parallel" network, utilizing unused fiber or copper connectivity. If planning a parallel security network, the existing cabling and connectors should be tested to affirm that they meet industry standards. Specific devices called "media converters" can connect copper/UTP network cabling to fiber optic links and vise versa. Technicians should test existing UTP network jacks for 10/100 Ethernet compatibility; some devices will only operate at one data speed or the other.

CHAPTER 15

Serial Communications and Ethernet

Many installations call for the inclusion or integration of access control communications and other monitoring and control functions, along with video and alarm monitoring. As the connection of access control equipment to Ethernet networks mirrors in many ways the technologies used to connect serial-based industrial processing and control systems, the following information is generally applicable to both types of installations.

How do these technologies communicate?

Remote access control panels and industrial monitoring modules communicate with their controlling head-end computers through serial data protocols such as RS-422 and RS-485. These interfacing standards were established by the Electronic Industry Association, and were developed to provide communications between devices for machine-to-machine communications.

RS-422 and -485

These data protocols transmit using asynchronous data communications, meaning that the data is sent one bit at a time, as shown in Figure 15-1.

The two most commonly used serial protocols are RS-422 and RS-485. In both types of networks, transceiver modules are typically connected using single twisted copper pairs and a separate signal ground conductor. Both RS-422 and 485 provide communications out to 4000 feet between devices, without the use of repeaters, which can extend that distance. This distance or the quantity of devices may vary depending on the equipment manufacturer.

RS-422 is typically used for point-to-point connections, while RS-485 offers the ability to connect up to 32 communication modules on a single network. Communications modules are addressed, typically using a "Com 1, 2, 3, . . ." scheme.

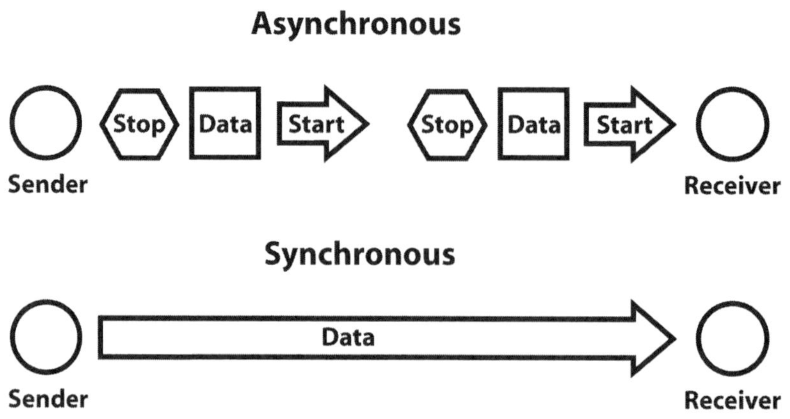

Asynchronous

Sender — ⬡Stop⬡ ▭Data▭ ⬦Start⬦ ⬡Stop⬡ ▭Data▭ ⬦Start⬦ — Receiver

Synchronous

Sender — Data ⟶ Receiver

Figure 15-1   Asynchronous.

At the controlling computer, the RS-422 or 485 cabling is connected to a hard-wired serial port, also called a "Com Port," on the desktop or laptop PC. This connection requires some form of interface to the PC, to convert the RS422 or RS485 to RS232 (i.e., the COM port on the PC), or directly into the PC via a specialized interface card. In the case of an access control system, the system management software will provide a selection for which "Com Port" is to be used to communicate over the RS-422/485 network.

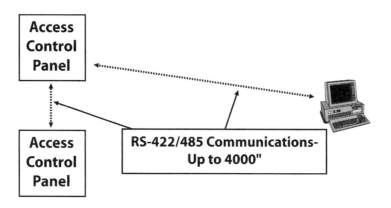

Figure 15-2   RS-422 and 485 networks can be as long as 4000 feet from one end to the furthest device.

Why change to Ethernet?

While these protocols are useful for relatively short distances, there are two reasons that it's beneficial to convert these protocols into Ethernet data. The first is the distance issue, as the conversion to Ethernet will allow devices to communicate over the WAN or Internet, allowing connections to devices across town or around the world. The other reason is that converting RS-422/485 signals into Ethernet provides the ability to control remote devices with standard PCs connected to the network, without separate connections for the RS-type communications. This allows the existing cabling to be used, leveraging network investments.

Ethernet converters/serial servers

To enable Ethernet communications with one or a network of RS-422/485 devices, a "serial server" is installed, which connects to the LAN/Internet and to the serial communications bus. Serial servers typically are available with 1, 2, 4, 8, or 16 serial ports. With the proper connections and IP address programming, the serial server becomes a node on the network and can be remotely accessed to download data, interrogate, or change the programming of the connected RS-422/485 device(s). If multiple devices are connected to the serial server, they are accessible using their "Com 1, 2, 3 . . ." types of individual serial addresses, once the serial server has been reached by the controlling computer. Serial server devices are also available for the RS-232 protocol.

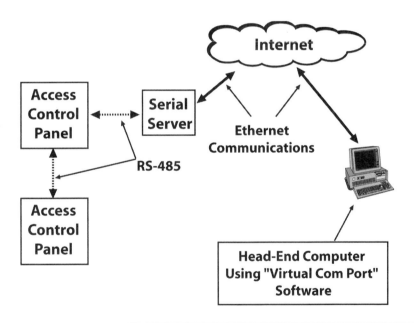

Figure 15-3   Ethernet/Serial access control communications.

SECURITY TECHNICIAN'S NOTE: Because they are in such widespread use, Ethernet devices tend to be much less expensive than other communication protocols. If serial communications such as RS-422 or 485 need to be sent over fiber optic links, it may well be less expensive to convert the signals to Ethernet, and then use UTP/fiber media converts to complete the communications path.

Head-end communications

At the head-end computer, "Virtual Com Port" software is installed, which associates a COM port connection with the specific IP address of the virtual server connected to the RS-422/485 network. The "Virtual Com Port" program converts serial commands and communications into Ethernet packets, which are transmitted over the LAN/Internet and processed by the remote virtual server, providing communications to the devices connected to the RS-422/485 network.

Serial tunneling

A second connection option provides the ability to establish constant two-way communication between one RS-422/485 device (or network of devices) and another, with or without the inclusion of a PC or intelligent device. In a manner similar to establishing an Ethernet "bridge," "Serial Tunneling" is accomplished using two serial servers that are programmed to constantly communicate with each other. Also called a "Nailed Down" configuration, this arrangement makes the Ethernet and cabling transparent to the serial communications flowing from one serial server to the other.

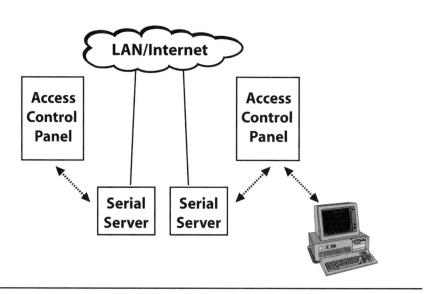

Figure 15-4 Serial tunneling.

This type of connection can be utilized to connect a remote RS-422/485 network to a monitoring computer, using a hardwire serial port on the computer.

Communications integration

Some security equipment manufacturers provide this "serial tunneling" capability within network video servers. Such devices can provide remote video, access control communications, two-way audio, and alarm inputs and outputs, all from a single networked communications device.

Figure 15-5 provides an example of how one security vendor's network server operates.

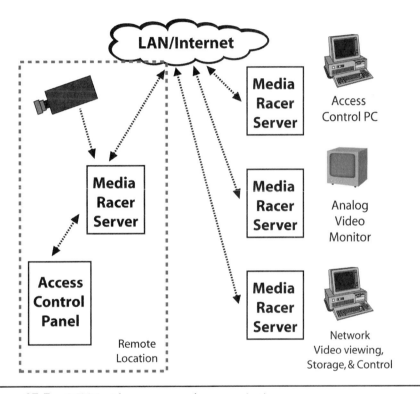

Figure 15-5 CCTV with access control communications.

As is seen in Figure 15-5, the Mavix "Media Racer" network servers can be addressed to transmit and receive access control information from one monitoring computer while transmitting video information to another. By using "serial tunneling," separate communication paths can be established for various types of signals.

Access control communication links

It's a wise idea to thoroughly investigate the potential options for connecting remote access control panels to a monitoring computer or head-end. The manufacturer may either recommend or build specific serial servers for its equipment.

Using the approved products is most likely a wise decision, as the vendor may choose not to support an installed system that is using non-recommended communication links.

Summary

Access control devices and some pan/tilt/zoom outputs are typically transmitted in either the RS-422 or RS-485 serial communication protocol. Devices using these protocols can be converted to Ethernet, which can extend signal transmission distances while leveraging an existing cable plant. Such communications can be either on demand, or set up as a constant "Nailed Down" communications link. Technicians should carefully investigate which specific devices the security equipment vendor recommends.

CHAPTER 16

Planning a Network Video System Installation

When considering a network video application, the installing company can face a wide variety of options in terms of equipment and software that can be used. To achieve a successful installation, security dealers must carefully consider bandwidth limitations, picture quality, desired monitoring and control capabilities, and potential future expansion of the proposed system. The following section will highlight potential choices of devices, and explain some general benefits or detriments of particular types of equipment.

AUTHOR'S NOTE: The following sections of this guide contain references to specific products from specific vendors. These references are used to illustrate the features and applications of security networking products. The inclusion of information about a particular product in this guide is not an endorsement by the author or the Security Networking Institute of a product's functionality in any particular installation.

The cameras

Either network-enabled cameras or standard analog CCTV cameras connected to network video servers can be selected to provide the transmission of video images onto the network. Below are the pluses and minuses of each technology.

Network camera benefits

1. Built-in Ethernet NIC's or Wi-Fi transceivers—This provides a smaller form factor, less equipment to install, and possibly lower cost than the analog camera and video server equivalent.
2. Wi-Fi Portability—Wireless cameras can be moved or relocated throughout the Wi-Fi network's coverage area, providing temporary surveillance options for the client.

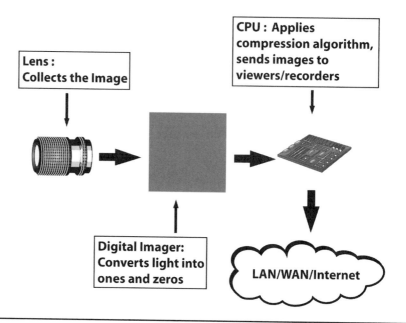

Lens :
Collects the Image

CPU : Applies
compression algorithm,
sends images to
viewers/recorders

Digital Imager:
Converts light into
ones and zeros

LAN/WAN/Internet

Figure 16-1   Functions of a network camera.

3. Built-in Web Server—Authorized users can access the camera(s) over the network using standard Internet browser software such as Internet Explorer or Netscape.

4. Quick Installation—Although not optimal, network cameras can be connected to any available Cat 5 jack using a patch cord, providing rapid installation.

5. Image Transmission Options—Some network cameras can be programmed to transmit images via email or to FTP (File Transfer Protocol) servers for storage and historical viewing.

6. Motion Detection/Alarm Inputs—Some network cameras have selectable video motion detection that can provide "alarm" images when motion is detected. Also, some cameras provide optional inputs for door contacts or other alarm activation devices that can trigger the storage and transmission of "alarm" messages.

7. On Board Image Storage—Some network cameras can store alarm images themselves, which can be viewed by connecting to the camera over the network.

Network camera concerns or limitations

1. Proprietary Compression Codecs—Due to their use of different compression/decompression schemes, network cameras from different vendors may not easily interface with each other at the viewing console or computer.

2. Picture Quality/Frames Per Second—Many network cameras are limited in terms of how many images per second they will transmit, and at what quality level.

3. Lens Selection—Many lower cost network cameras have fixed lenses and provide limited options for focusing, backlight compensation, and other video image issues.

4. Location Limitations—Many network-enabled cameras have been designed for indoor use only.

5. Recording Options—While a network camera may transmit images to an FTP site, retrieval and viewing of these images can be quite cumbersome for the system operator. An auxiliary software set may be needed to provide practical video storage and historical viewing.

6. Maximum Number of Cameras—Although there are exceptions, most network camera products are designed so that a maximum of four individual images can be viewed simultaneously on a single computer screen.

7. Limited Compression Options—Some network cameras use proprietary compression schemes that provide no options for the installing dealer. File size is dictated by the "scaling" of the image, with 640×480 providing the best image and biggest file size, while 160×120 provides a degraded picture with a much smaller file size.

Network cameras—the bottom line

Network cameras are the right choice for indoor residential and light commercial projects where there is little possibility of the system growing beyond four cameras. Wi-Fi-equipped cameras, such as the Veo "Observer" and the SOHO Wireless Internet Camera, provide quick installation and flexibility of camera location. For example, parents can plug the Veo Observer into an outside AC outlet, aim it at the children playing in the back yard, and view the scene from their Wi-Fi-equipped laptop, which they can carry from the kitchen to the bedroom, or anywhere within the Wi-Fi network's coverage.

Low cost network cameras may have difficult-to-use recording and historical viewing interfaces or may require specific software to view them, although some are better than others. Network cameras are simple to install and sell, as any Ethernet or Wi-Fi-equipped laptop can be used to demonstrate the camera's capabilities. Just plug in the wired camera, or program the Wi-Fi camera and laptop in "ad hoc" mode, and the salesperson can demonstrate the camera right before the client's eyes.

Analog camera and network video server benefits

1. Connect to Existing Cameras and Hardware—Video servers can connect existing cameras or quad viewing equipment to the network and the Internet.

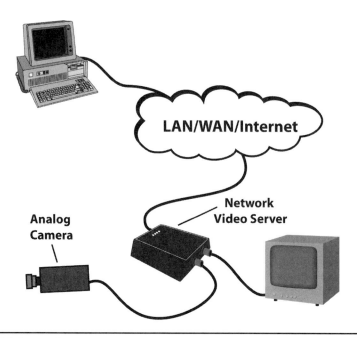

Figure 16-2   Analog CCTV on the network.

2. Use any Analog Camera—Indoors, outdoors, p/t/z, all can be connected to video servers to transmit their images, and to receive control information.

3. Large Scale Networks and Systems—Video server systems can be used to connect hundreds of cameras into a single monitoring system.

4. Audio, Alarm Interfacing, other Outputs—Some video servers provide two-way audio, alarm input monitoring, and control output functions so that a single server can provide video, two-way audio, and control, all from one device.

5. Local Monitoring Capability—Some video servers are equipped with a "pass through" video connection, allowing the video signal to be transmitted over the network as well as viewed on a local CCTV monitor.

Analog camera and network server concerns or limitations

1. Cost—An analog camera and video server will likely cost more than an equivalent network camera. However, the video server generally provides more features and options than a similar network-enabled camera.

2. Proprietary Compression Codecs—As with the network camera, video servers can also utilize proprietary software for compression and decompression of video signals. As such, it may be difficult to interface video servers from one vendor with video servers from another at the viewing station.

Analog camera and network server—the bottom line

Network servers are the correct selection for large integrated systems requiring the connection of multiple cameras, access control, audio, alarm inputs, and control outputs. Existing CCTV systems can be connected to the network with video servers, pushing the images over the network while retaining local viewing on the existing CCTV monitor.

> **SECURITY TECHNICIAN'S NOTE:** The camera, lighting, and lens determine the quality of a video image. It's important to remember that putting video onto a network will not improve the images viewed or recorded. Based on the type and settings of the compression within the network camera or encoder, it's likely that network video image quality will be reduced when viewed, in comparison to a standard analog camera arrangement.

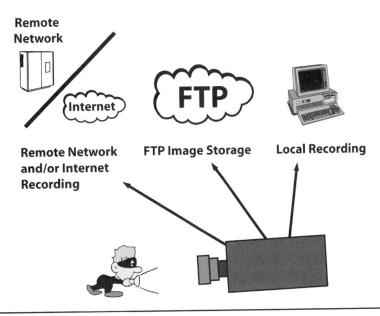

Figure 16-3   Rrecording options.

Monitoring and recording options

After the selection has been made regarding the types of camera(s) or video servers to be installed, consideration must be given to how the camera's images are to be viewed, controlled, and recorded. This is a critical issue, and is fully explored in later sections of this guide. It's important, when planning which camera technology is to be used on a particular installation, that the selected devices will provide the quality and quantity (fps) of images needed to meet the client's viewing expectations. It is also important to confirm that the video compression codec used by the camera or video server will properly function with the recording method selected.

Summary

Security products manufacturers are producing a wide variety of network cameras, video servers, and video decoders. Security dealers need to investigate the video transmission, viewing, and communications options of specific products to determine their usability for a specific application. Network cameras provide an integrated solution, while video servers can be connected to existing (or new) traditional analog security cameras.

Video Compression Technologies

In the planning and installation of a networked video system, compression technologies play a critical part in reducing the size of files for transmission over the network and storage. The following is a simplified explanation of this complex topic, which will provide the reader with a clear understanding of why video compression is utilized and how it works.

Streams of images

The first issue to understand is what "video" is, in terms of viewing moving images on a television or computer screen. "Video" as seen on a CCTV monitor, TV show or movie, is a rapidly changing series of still pictures or images. The typical video signal is composed of between 25 and 30 frames (images) per second. This is the "frames per second" (fps) rate which we as viewers are accustomed to seeing when we view a television screen. With CCTV systems, the individual frames can be copied from the DVR or recording server to provide evidence for investigative purposes.

In a network video system it may be necessary to transmit 25–30 fps over the network, and (most likely) to provide storage for these images. As will be detailed later, these requirements must be carefully considered when designing a system to ensure that an adequate video signal is received and that enough storage space is available to retain the amount and quality of video desired by the client.

Basic television

Some understanding of how standard television images are presented is important to comprehending how video compression can affect the quality and transmission of network video signals.

The screen of a color television tube is coated with a layer of red, green, and blue phosphor elements. Electron "guns," one for each color, shoot electrons at the phosphor elements, causing them to glow briefly. Red, green, and blue are used, as they can be mixed together to recreate virtually any color, from white to black and all hues in between. Each cluster of one red, one blue, and one green phosphor element is called a "picture element," or "pel." The intensity of the electron gun's electron stream causes the elements to glow at various brightness levels.

To present a complete image, the electron guns "light up" the phosphor elements by "painting" the screen in lines. The lines for each individual image are painted in two complete passes of the electron guns, which paint the image in alternating lines. The completed image is composed of both sets of lines. This dual line concept is called "interlaced" video, and is common to television and CCTV systems.

Resolution lines

The number of phosphor element pels on a television screen is called the resolution, with more pels providing a more detailed image. The pels are arranged on the screen in horizontal lines, with a specific number of pels per line. The aspect ratio of standard television is the comparison of the width of the viewing screen to its height. The standard aspect ratio of television displays is expressed as 4:3, with the horizontal being longer than the vertical.

The National Television Standards Committee (NTSC) standard for American-type video transmission specifies 640 horizontal pels, and 480 vertical lines. This format is also used by most analog CCTV cameras and monitors.

Composite video

Initial television systems were black and white, with the electron gun activating the phosphor elements for brightness only. In order to display color images, while retaining the ability of black and white television sets to receive transmissions, color TV signals include the signals for color phosphor activation, which are transmitted along with the "brightness" information. The resulting signal is called "composite video."

From analog to digital

Although some standard CCTV cameras are called "digital," the output of the camera is an analog signal, suitable for viewing on standard CCTV monitors. As described in Chapter 16, while a CCD imager receives the light coming through the camera lens and outputs digital information, this stream of ones and zeros is converted into an analog output for standard video use.

In order to transmit a video stream over a network, the individual images must be converted from their analog state to binary computer language. This process is called "digitizing."

Each pel signal is sampled by the digitizing software, which converts the brightness and color signals into a binary data stream, as seen in Figure 17-1.

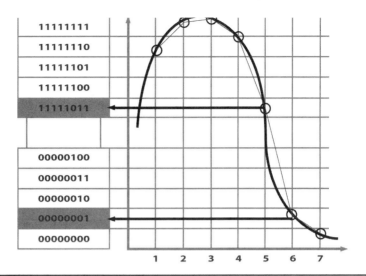

Figure 17-1 Bit Sampling. Each time the signal is sampled, the amplitude is different. Each sample falls into a zone represented by an 8-bit level. 8-bit sampling provides 255 potentially different values for each sample.

The digitizing of video signals is performed based on an industry standard designated "ITU-R BT.601," developed by the International Telecommunications Union.

Lossy and lossless compression

After the analog signal has been sampled and converted into a digital binary code scheme, the compression process begins. Compression of video signals can be achieved by a number of methods. Some methods are "lossy," achieving file size reduction by eliminating some information from the image, while other methods are "lossless," as they retain all original image information. These compression technologies are often combined to produce the best possible picture with the smallest file size. Popular compression formats such as MPEG can use both techniques to reduce file size, and users can set certain parameters within the compression "codec," which is a computer term for any technology that compresses and decompresses data.

While security installation companies will generally be primarily concerned with overall picture quality, frame rate, and bandwidth requirements, the issue of "lossy" versus "lossless" compression is important to consider. If the stored video images are to be potentially used for evidence, lossless compression is a better option, as no parts of the images are removed.

SECURITY TECHNICIAN'S NOTE: "Lossy" compression equals lost data. If a video image is highly compressed, details of the viewed scene may be lost in the compression process, and cannot be reconstituted.

Decimation

Typical compression codecs will divide the image into 8 by 8 blocks of pels, and will separate the black-and-white brightness component, called the "luminance block," from the two color components, called the "hue" and the "saturation" components. These two color components together are called the "chromatic blocks." Each of these components is processed as an individual 8 by 8 grid of data.

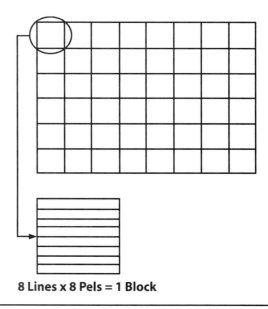

8 Lines x 8 Pels = 1 Block

Figure 17-2   8 by blocks. The compression codec samples and processes image information in individual blocks for data transimission.

The human eye is more sensitive to dark and light than to slight color changes. Compression codecs can take advantage of this situation by discarding a portion of the color components. This process is called "decimation," and is included in the concept of "scaling."

"Horizontal decimation" is the process of discarding every other line of the color components in the chromatic blocks. As there is twice the amount of chromatic data as luminance data in a set of blocks, this process alone can greatly reduce file size.

More compression can be achieved by only keeping every other horizontal and vertical chromatic data bit. This is called "horizontal and vertical decimation."

Scaling

Reduction of the data to be compressed through decimation, and the number of blocks by which the uncompressed image is to be processed, is performed within the process of "scaling." Usually selectable through software, the level of scaling determines the quality and resolution of the image after passing through the codec.

Scaling defines and adjusts the amount of information that will enter the codec for further processing. It's important to remember that once the picture is scaled, whatever information is lost through the decimation process is gone forever. The compression process will not improve the picture, and likely will degrade it further.

Spatial redundancy

The goal of a compression technology is to reduce the file size. In the scaling process, the image is divided into a certain number of squares. If the image is of a person in front of a white wall, many of the blocks in the image will be of the white wall. Instead of transmitting each and every white colored block, compression codecs will transmit one white block, with instructions as to where the multiple blocks of white should be placed on the reconstituted image grid of the viewing display.

Temporal redundancy

Another method of video image file size reduction takes advantage of redundant segments within a stream of pictures. If a politician is giving a speech on television, the background stays virtually the same as the politician proselytizes. Compression codecs take advantage of this phenomenon, and will only transmit the changes in successive images. As this information is redundant, and will be reconstituted at the receiving end, the quality of the video is not adversely affected by the removal of temporal redundant information.

Further file size reduction

Once the image has been scaled, various complex mathematic programs are used to further reduce the file size. These are typically standardized, based on the type of compression used.

JPEG compression

The Joint Photographic Experts Group developed one of the first compression systems, called JPEG. Using many of the concepts mentioned above, JPEG is a widely used codec for the digital storage and transmission of photographic images. As a video stream is a series of individual images, JPEG has been used as the basis for subsequent codec technology, the most commonly used being the MPEG compression technologies. As a note, if JPEG is used to provide a video stream transmission, it may be termed MJPEG, just to provide more confusion in a sea of acronyms.

MPEG

The Motion Picture Experts Group, called MPEG, has developed standards for the compression of video images. These standards are based on a JPEG type of compression of some of the individual images, while applying a "lossy" form of compression to subsequent images, providing a greater reduction in overall file size.

Greatly simplified, here is how MPEG works. Reference images called "intra" or "I-Frames," are compressed using JPEG-like methods. Subsequent frames called "Predicted" or "P-Frames" consist of any changes that have occurred in the I-Frame to which the individual P-Frames are referenced. Another type of frame, called a "bi-directional predictive" or "B-Frame," references two I-Frames, providing reverse viewing capability.

These frame types are transmitted in a format called a "Group of Pictures," or "GOP." The typical GOP will consist of an I-Frame and twelve to fifteen combined P and B frames. If the image is being transmitted at 30 fps and the GOP is selected for fifteen total frames of all types, two GOPs will be transmitted each second.

As individual frames may take different paths across a network or the Internet to reach the viewer/recorder, there will be a small amount of lag time or "latency" between what the camera is viewing at a particular point in time and what is appearing on the remote viewing screen. The receiving codec will reassemble the received frames in the proper order and display them.

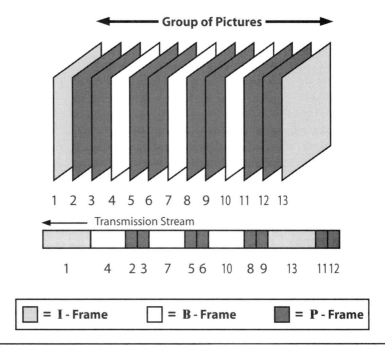

Figure 17-3   I, B, and P frames are transmitted out of order to the receiving computer, which uses the I and B frames to interpret and display the "lossy" P frames is a group of pictures (GOP).

MPEG2 is the most commonly used format for video compression and is the standard for DVDs, satellite television, and cable TV set top boxes. For network security video, many vendors are supplying their cameras and receiving software and/or hardware with some variation of MPEG2 compression.

Typical video compression algorithms used in networking

Unfortunately, a wide variety of compression technologies are currently provided in network security cameras and video servers. Often these codec sets are proprietary in nature, meaning that a network camera from vendor "A" cannot be viewed from the PC software supplied by vendor "B." This can force the security installation company to standardize on a particular product line for all video streams being used in a specific installation.

This will become more problematic as the installation of network security video moves forward. For example, consider a client "C," who has forty-five locations being remotely monitored using vendor A's equipment. Another client, "D," is in a similar business, and has eighty-five locations being monitored using equipment and software from vendor B. If client D buys out client C and wishes to consolidate both monitoring operations, there is a strong possibility that video transmission equipment will have to be changed out at many locations to achieve the desired results.

If planning to integrate video streams emanating from different vendors' products, the careful security installation company will test the different video feeds and verify their functionality with a particular receiving/recording software or hardware set.

Compressing the future

Video compression is a key element in the successful transmission and storage of security images. It is important to understand how it works, what adjustments can be made to increase or decrease file sizes and image quality to fit bandwidth limitations, and how the selection of devices to be installed can effect image viewing and recording.

Video Bandwidth Controls for Shared Networks

The following section details some of the choices available to limit the bandwidth requirements for the transmission of video images over an IP network. This discussion is very important if the cameras/video servers are sharing a network. To a lesser extent it is still a consideration if the IP imaging devices are being connected to a parallel or private network.

Why bandwidth control is needed

When connecting video transmission equipment to existing data networks, a detailed knowledge of bandwidth needs and how to control them will help the security installation company best work with IT personnel, who are justifiably concerned that the introduction of video signals will degrade their enterprise network performance.

Network bandwidth availability and variables

Which computers are using a network at any given time, and how much bandwidth they are consuming, are constantly shifting variables. While a network may be quite busy during a typical business day, during the night or on weekends there may be little or no enterprise traffic.

Where video-imaging bandwidth needs differ from enterprise communications is that video streams will require a fairly steady amount of bandwidth to transmit enough frames per second, at a high enough quality, to achieve acceptable viewing and recording of images. Generally, this constant video stream will be flowing on a 24/7 basis, as most video security systems are configured for constant monitoring and recording. So we can consider the video streams to constitute a "baseline" of consumed bandwidth on the network, leaving the remainder for the fluctuating needs of enterprise communications. (Of course, the IT manager doesn't consider the enterprise communications to be of secondary importance to the video streams!)

Networks are constantly changing and growing to best suit the needs of the users connected to it. Security dealers must consider future expansion of network systems, both from the security device perspective and also that of the enterprise.

Will the addition of ten or twenty more cameras clog the network? What if the client adds twenty more enterprise network users? What about one hundred? Potential growth of the number of devices behooves security dealers to configure their video systems to use a minimum amount of bandwidth, even if there appears to be plenty of room in the existing network.

Options for bandwidth control

The number of ways that an installing company can tailor bandwidth usage will be limited by the feature set included in the IP cameras or video servers selected for the particular job. The following discussion provides some details regarding various types of bandwidth-limiting options; not all products will have the complete suite of options. To further confuse the issue, the same option may be called by a different name, and be accessed differently in various products.

The three generally available selections for bandwidth limitation are frames per second, scaling, and compression percentage. It is the combination of these factors that will increase or decrease bandwidth usage; changing any one, or all, of the selections will affect bandwidth requirements.

Frames per second (fps)

One of the primal concepts of CCTV is that the only acceptable video signal is "real time," usually considered to be 30 fps. Although this is the standard for old technology analog CCTV, this concept needs to be reconsidered in light of bandwidth requirements and digital storage space concerns for the successful use of network video products.

A simple method for reducing network video bandwidth usage is to reduce the fps rate being transmitted. As each frame is a complete or partial image, reducing the fps can dramatically cut the bandwidth needed for that particular camera. The great majority of current IP cameras and video servers provide a simple selection for the maximum fps that a particular device will transmit onto the network.

Many security dealers have experienced good success using 15 fps, which generally provides enough motion for viewing and recording needs. Rates as low as 7 fps are often used when a sizeable number of network video images need to be recorded.

Frame rate settings can be flexible, with some cameras in a system being set for higher rates while others transmit at slower fps. When using analog cameras connected to a DVR, recording rates can temporarily change from a slow fps setting to "real time" 30 fps after being activated by software video motion detection or an auxiliary trigger input.

Running at slower than 30 fps may cause a potential inadequacy in performance and possibly a client relation problem, as the customer may feel that the system isn't providing "real time" video. *Security salespeople and installers who are providing network video solutions need to become comfortable with reduced frame rates and represent the system to the client in a realistic way, to best avoid customer disappointment.*

Smart security dealers can set up a video server and camera at their office, and use it to demonstrate different fps rates for potential clients. Such demonstrations can be performed from an Internet-connected PC at the client's location, providing a "real world" example of the power of remote IP security imaging.

If "real time" video is a requirement for a particular job, 25 to 30 fps can be achieved by reducing the sampling rate (scaling) and increasing the compression percentage, while using a reasonable amount of bandwidth. Consider the following simplified equation:

Video Bandwidth = (scaling × compression rate) × fps

Reducing any of the components will reduce the overall bandwidth needed.

Adjusting the fps is a simple selection that retains the image quality while positively impacting the bandwidth issue.

> **SECURITY TECHNICIAN'S NOTE:** Security company personnel need to adjust their thinking with regard to frames per second when planning and installing network video systems. While 30 fps is typical for analog cameras, 15 fps is a reasonable video stream over a LAN, although, most Internet connections may provide 3-5 fps when viewed remotely. Security system salespeople should use 15 fps to demonstrate network video to their clients, to give them a realistic sense of how the video will look when the system is installed.

Image scaling

In general terms, "scaling" is a programmable selection in an IP camera or video server that defines how many lines of pixels of each image will be compressed and transmitted over the network. Figure 18-1 shows the typical settings for a video server.

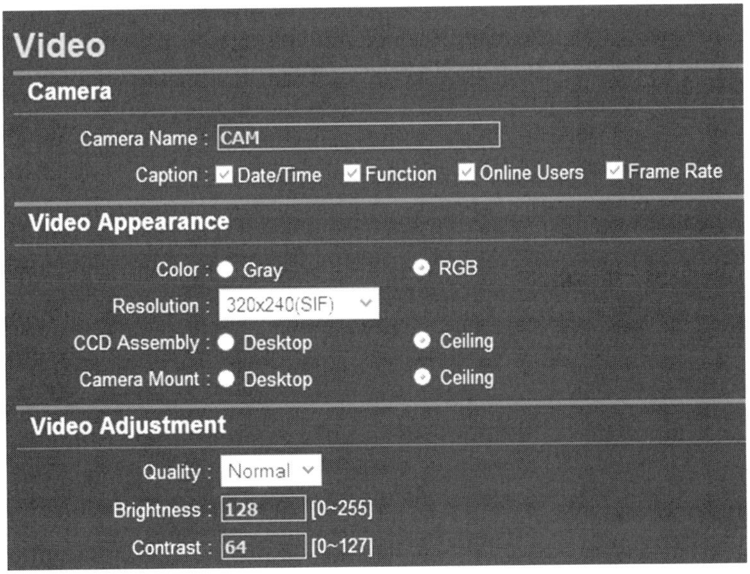

Figure 18-1 Blue Net Video settings.

This video server provides scaling options in the "Resolution" selections. While 640 × 480 provides the best image and the largest file sizes for transport, this particular device is set for 320 × 240. Choosing one of the lesser scaling settings greatly reduces file size. By reducing scaling settings, technicians also reduce overall compression/decompression time. That enables images to be displayed virtually instantaneously. In contrast, a slight delay is experienced when 640 × 480 scaling is used.

Scaling, fps, and Internet connections

When viewed over the Internet, maximum frame rates will be dictated by the available upstream bandwidth of the client's Internet connection. Adjusting the scaling downward will increase the number of frames that can pass through the connection to the remote viewing computer per second. Remember that "upstream" is data being sent and "downstream" is data being received. There is typically quite a difference between the two, with the downstream capability usually much larger than the upstream when connected to a typical DSL or cable ISP service.

Table 18-1 shows some examples of the potential number of frames that can pass through a typical DSL connection providing a 128k upstream bandwidth, at different scaling levels.

TABLE 18-1   The Potential Number of Frames That Can Pass Through a Typical DSL Connection

Image Resolution	Range of Image File Sizes	Frames Per Second @128 kbps
160 × 120	32–48 k	3–4
320 × 240	64–96 k	1–2
640 × 480	320–480 k	0

Notice that as the scaling level increases, the number of potential frames per second decreases. It's important to know that in the example, the 640 × 480 scaling level will transmit images, but they will be received at a rate of one frame every few seconds.

Compression percentages

Often an IP camera or video server will have an "image quality" selection, with descending choices such as "Superfine," "Fine," "Normal," and "Low." These options relate to the percentage of compression that is applied to the individual images, with the highest quality being the least compressed. Less compression equals better image quality and motion capture, but means larger file sizes, slower transmission speeds, and larger storage requirements.

Compression method selection

Different manufacturers have developed their IP network cameras and video servers to work with specific compression methods. Typically a product will provide either JPEG or its motion adjunct MJPEG, or MPEG. While JPEG/MJPEG transmits whole and complete pictures, MPEG ships some complete and some partial frames, which are referenced to the complete pictures.

In general, MPEG will provide smaller file sizes, while JPEG provides complete images that can be individually copied from the storage disk for transmission and viewing.

It is important in the product selection process for the dealer to ascertain that the compression codec provided is compatible with the planned storage and viewing equipment and software.

Variables of bandwidth usage

The amount of bandwidth that a particular camera/video server needs to provide adequate quality video can vary based on network factors and the type of images being viewed. Motion in a camera's image requires more bandwidth to transmit than a still image, as the images are more complex. Cameras viewing a busy scene, or a camera that is programmed for preset pan/tilt/zoom functions, will require more bandwidth than a stationary camera viewing a relatively "quiet" scene.

Good, better, best

For the security dealer, it is a natural choice to program systems to provide the best possible image for viewing, at the highest levels of fps, image quality, and scaling. After all, we're security technicians, and want our systems to perform to their utmost. However, when sharing the network cabling and transmission equipment with the enterprise, the astute security installer will most likely want to reduce image quality and fps, to leave as much room on the shared network as possible for enterprise communications. Remember, the security system and the enterprise must *both* function properly, and the IT personnel can easily cause service problems for the security dealer after the installation. They are at the client's location every day, and the security company has to roll a truck to respond to a problem.

Calculating bandwidth needs

IT managers will be much more amiable to the concept of sharing bandwidth with network video systems if the security company can provide detailed estimates of bandwidth requirements.

Let's take a look at some sample bandwidth calculations in Table 18-2. These figures, provided by BlueNetVideo, are based on MJPEG compression ratios.

TABLE 18-2   **Bandwidth Calculations**

Pixel/ Frame	Bit/ Pixel	Bit/ Frame	Frame/ Second	Total Bits/ Second	Bandwidth Range in Mbps-Fastest-Best	Typical Bandwidth
160 × 120	16	307,200	30	9,216,000	.31–1.84	.46
320 × 240	16	1,228,800	30	36,864,000	1.23–7.37	1.84
640 × 480	16	4,915,200	12	58,982,400	1.97–11.8	2.95

Let's walk through one of the examples in Table 18-2. Using 320 × 240 scaling and a sampling rate of 16 bits per pixel, the size of each complete frame is 1,228,800 bits, or 1.228 Mbps. Multiplied by a frame rate of 30 per second, the bandwidth requirement *before compression* is 36.864 Mbps. Remember that a typical Ethernet LAN provides a maximum of 100 Mbps per segment, so a single camera at these settings would consume one-third of the available bandwidth by itself without the magic of image compression.

Now let's apply various levels of MJPEG compression to the 320 × 240 30 fps image stream. Using the "Best" compression percentage, which provides a 5:1 file reduction, the data stream has been reduced to 7.37 Mbps (36.864 divided by 5). Selecting the "Fastest" compression percentage will reduce file size to a 30:1 ratio, reducing the bandwidth requirement to 1.23 Mbps. Actual bandwidth needs will also depend on the image viewed; movement within the image requires more bandwidth than when no movement is present.

Reviewing this table provides an indication of how compression ratio settings will increase or reduce file size. What cannot be described in mathematical terms is what the quality of the image will be, and how clearly movement will be viewed. These factors can only be determined by actually viewing the camera over the network, determining the minimum quality parameters, and setting fps and compression ratios to provide the necessary quality while using minimal bandwidth.

Black magic

Because of the wide number of variables, there are no hard and fast rules for the settings that determine the bandwidth requirement for a particular camera or video server when transmitting over a shared network. It is best for the installing company to determine a bandwidth "budget" that provides a calculated high/low Mbps usage calculation for each connected device. These calculations can then be presented to the IT manager, providing him with the anticipated requirements for the new devices being added to his network.

Summary

Placing one or more IP-enabled video security devices onto a network will require a constant amount of point-to-point bandwidth to pass the video images. Video bandwidth can be controlled by the compression type selection, with MPEG transmission providing a smaller data stream than a comparable JPEG stream. After the compression selection, reduction of the frames per second rate, image resolution, and compression adjustment can reduce bandwidth needs. When bandwidth usage is limited using these methods, often image and viewing quality will be lessened.

CHAPTER 19

Video Control and Recording Options

When considering the installation of an IP-enabled camera or video server, issues regarding how the image can be controlled and recorded must be addressed. The following section will detail the available options for image control and the various recording/image transmission capabilities that are often present in typical network-enabled cameras and servers.

Image control

Typical IP-addressed cameras and video servers offer the ability to control brightness, saturation, and other aspects of the image quality itself. The "Media Racer" series of video servers from Mavix will be used as an example in this section.

While most networked cameras and video servers offer similar control features, there is little uniformity between vendors in terms of where in the programming certain selections reside.

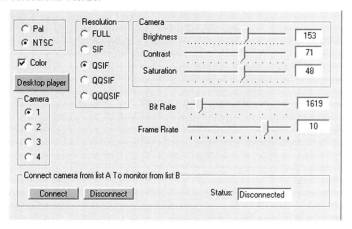

Figure 19-1   Mavix Video Configuration Screen.

The Media Racer product provides all image control selections on a single screen for easy location and manipulation. The "Resolution," selection provides a full range of scaling, while the "Brightness," "Contrast," and "Saturation" settings can be adjusted with the slider controls. One very useful feature of this product is that the image being manipulated can be viewed on the same screen as the selections, so that the impact of any changes can be readily seen and adjusted.

Below the image adjustment options, the "Bit Rate" and "Frame Rate" selections provide a powerful set of tools for bandwidth control. These two selections are relational, in that increasing the frame rate while leaving the bit rate constant will reduce the quality of the images but will leave the overall bandwidth used the same. Conversely, increasing the bit rate with a constant frame rate will increase the quality of images. Because of these bandwidth control features, the Media Racer products can transmit usable video images to low-bandwidth devices such as IP-enabled cellular telephones and personal digital assistants ("PDAs").

Camera movement

If a network-connected camera has pan/tilt/zoom capabilities, there are two primary ways in which camera and lens movements can be controlled. For products such as the Panasonic MV-NM100, pan and tilt movements can be controlled by authorized users connected via computer to the camera's web server.

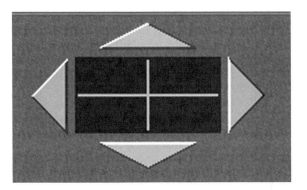

Figure 19-2 Panasonic Pan Tilt Screen.

Clicking on any of the arrows will move the camera in that direction; clicking anywhere on the center grid will immediately move the camera's aim to that spot. Panasonic includes a similar control interface on its DVR products, controlling analog dome cameras made by the same vendor.

If using an analog pan/tilt/zoom camera, most IP-enabled video servers have the capability to provide control and preset locations, using a similar interface to the one illustrated above. To ensure compatibility, installation technicians should confirm with the video server supplier that its product will provide the proper protocol for communications to a particular camera.

As IP networking grows in the electronic security field, manufacturers are producing more and varied types of addressed cameras, including p/t/z cameras with networking connections. Such cameras will typically be controlled over the LAN or Internet by connecting to the camera's web browser and inputting movement commands. As described below, video management software programs are available that can consolidate and automate the controlling of p/t/z commands for large numbers of cameras.

> **SECURITY TECHNICIAN'S NOTE:** Network "latency" (the time it takes a data packet to travel from point A to point B on a network) can cause network-based p/t/z control operations to "swing by" a target viewing area. As the operator clicks the "right" or "left" input, he or she will often click too many times, as the movement of the camera is slowed by network latency.

Image storage, transmission, and recording

Although the presence of a camera itself can provide some level of deterrence to criminal activities, in many cases the primary role of surveillance video systems is to provide "after-the-fact" information, allowing users and authorities to review who did what, where, and when. IP network cameras and video servers provide selections from a variety of stored image options, including "in the camera" storage, email transmission, and File Transfer Protocol. Software packages are available that provide Network Video Recording (NVR) capabilities, recording stored images onto hard disk drives. The follow section details these typical image storage options and their uses.

In the camera

Some IP-enabled cameras, such as the Panasonic MV-NM100, provide storage of a number of images within the camera itself. Storage of images is typically triggered by the activation of pre-set video motion detection, or an external input such as a door contact or motion detector that is wired to the camera. Products with internal storage will typically have settings for the number of pre- and post-alarm images that are to be stored for a single event. Once these images have been saved in the camera, they are accessible to authorized users through the camera's web server function.

Although this can be a very useful function, storage space within an IP camera is generally quite limited. The Panasonic camera mentioned above will store approximately sixty images, depending on the compression ratio and scaling settings for the camera, with better quality images producing larger file sizes.

For this "in-camera" storage to be utilized, images must be viewed quickly after an event takes place. Most IP cameras providing this image storage option will overwrite the storage file space when a new alarm is activated, partially or completely erasing previously stored images.

Email

Many IP video servers and cameras equipped with video motion detection and/or external alarm trigger inputs offer the option of email messaging, sending out a pre-programmed message—with or without an attached image or set of images—to previously programmed email addresses. If the camera is connected to the

Internet, such email messages can be transmitted to any Internet-connected device that has its own email address.

The potential uses of email image transmission provide an excellent example of the ways in which electronic surveillance systems can utilize other evolving technologies to increase and improve client notification and awareness of events.

For example, if there is a network camera installed in a retail store and it is triggered to transmit an alarm email message, the message can be sent to the owner's home email address. But if the owner's child is on the computer at the time of the event, the email message might not be seen by the proprietor until the next day or later. Many cellular telephone companies have upgraded their networks and can provide email and image transmission to wireless telephones. This enables the email alert to be programmed to transmit to the owner's cell phone, alerting him to an event at his store. The same scenario can also apply to a roving guard at a large facility, or other personnel who should be immediately informed of an event.

File transfer protocol (FTP)

Commonly used for file transfer and storage on the Internet, many IP cameras and video servers offer the option of FTP storage of video images. FTP storage can be located anywhere on either the LAN or on an Internet-connected server, which can be in any location.

FTP servers are commonly accessed using web browser programs. An FTP address will read something like ftp.universalsalvage.best.com. Typically, FTP sites require user authentication, though some allow anonymous connections. When the server has been reached, the user must input a valid username and password to be allowed access onto the site. Once in the website, files are arranged in directories and folders, and can be either "uploaded" onto the FTP site, or downloaded. "Downloading" means that a copy of the file is made and transferred to the user's PC. FTP servers are generally used to provide a separate location, different from a user's primary network, where files can be shared between authorized users.

Those IP cameras and video servers that provide FTP capability can be programmed to automatically connect to the FTP server, enter the appropriate username/password, and place video image files into specific directories within the FTP server. File names are automatically generated by the transmitting camera, usually with some permutation of the date and time of the image's creation.

FTP image transfer can be set to happen on a timed basis, and scheduled by the day of the week. Alarm images can also be transferred to the FTP server. The options and capabilities of FTP will vary from one IP camera to another; some offer no FTP capabilities.

FTP image transfer provides a powerful security feature in that video images can be stored remotely from the camera's location, preventing criminals from destroying image evidence, which can easily occur when video is recorded on site.

One of the drawbacks of FTP image storage is that, with few exceptions, FTP images will be stored as single images only. This can make finding the right image for evidentiary purposes difficult, as the security director may have to wade through hundreds of uploaded images to find the one that is important. FTP image storage is also limited to devices using complete image compression algorithms, such as the MJPEG or JPEG codecs.

FTP provides an effective, albeit somewhat awkward, image transmission and retrieval method.

Web browser vs. software program control

Although most network-enabled cameras and video servers can be controlled by using a web browser, this method of image and camera movement control is cumbersome and slow. Authorized users must either click a "favorite place" listing on their web browser software or individually input the IP address of each camera to allow changes to be made. Web browser access also can be a limiting factor for viewing and recording purposes, as usually only one camera can be viewed and controlled at a time. Web browser control of IP cameras or video servers should be limited to a small population of such devices, perhaps no more than two.

Network video recorder (NVR)

Software programs are available that provide full featured viewing and camera control, as well as storage and retrieval of network video images from multiple IP cameras and/or video servers. Once installed into the primary monitoring computer, these software packages will provide automatic or pre-programmed communications with network or Internet-connected cameras. This eliminates the need for authorized users to sign onto each camera or video server individually to view or control those devices. These programs can also direct the storage of image files onto digital media, such as hard disk drives, which can be located either within the primary monitoring computer or remotely at another location on the network or Internet.

These combinations of software, hardware, and features for video image manipulation can create some confusion, as various manufacturers will use different terms to describe their products' interactions and functions. Some vendors have software that provides image viewing, alarm interfacing, and camera control, while providing a separate program for recording and retrieval of stored images. Other vendors combine all of these functions into a single software program.

A network video recorder (NVR) is generally considered to consist of a software program that directs video storage and retrieval, a dedicated PC, and hard disk drive (HDD) arrays. The latter should be of sufficient size to store weeks or months of video that can be retrieved for review.

A network video management program (NVMP) provides camera display and control features along with the recording capabilities of an NVR. NVMP programs can be quite inexpensive, and provide extensive viewing and recording options. A "guided tour" of the ReCam NVMP software is included in Chapter 28.

Terminology associated with these devices can sometimes be quite confusing. Installation companies should carefully investigate the capabilities of software products that they are considering for a particular project.

Compression compatibility issues

A key and potentially complex issue for security professionals who are selecting NVR or NVMP products is the compatibility of field devices with the chosen software package.

While some software packages can interact with IP cameras and video servers from a variety of manufacturers, others will only function with "mated" devices from the same vendor. There are three areas where incompatibility can occur—initial sign-on, compression codec, and control of p/t/z functions.

A problem in any of these areas can disrupt control capabilities and communication with the remote video device. Until our industry standardizes on these issues, which is admittedly wishful thinking, incompatibilities between equipment and monitoring/recording software will be present.

Multiple image transfer options

Based on the IP-addressed cameras, video servers, and software selected, systems can be configured to provide multiple video streams, each intended for different purposes. For example, using an IP camera with internal video motion detection, email, and FTP image transfers can be programmed to go to select Internet devices, such as a guard's PDA. Regularly scheduled FTP images, meanwhile, could be stored on an off-site web server. If the monitoring system is equipped with NVMP software, additional video images can be stored on one or more separate servers, which could be at different locations on the Internet or network, enabling backup image storage.

Sensory overload

Video imaging transmission and storage for large-scale systems must be carefully planned so as to avoid overflowing available storage capacity. Care should also be taken to prevent hundreds of daily emailed images that could be generated by a camera set for motion detection and email transmission.

Summary

Different video security devices will offer various methods of image control and recording options. Some devices offer self-contained image storage, while FTP and email can also be found in some products. Network Video Recorders (NVRs) can provide video storage and reviewing, while an NVMP (Network Video Management Program) can provide integration of security video images with alarm and access control information. Some products provide multiple output options that can be tailored to provide specific video feeds for specific purposes. It's important to note that there is currently little cross-platform compatibility between competing vendor products; to view the video from vendor "A's" encoder will often require Vendor "A's" software.

CHAPTER 20

Powering Devices

One of the wonders of the world of networking is how many different devices from different manufacturers can all be programmed to communicate with each other on the same LAN or over the Internet. Thanks to the established standards, equipment and software designers build products and software platforms that can all work together in a harmonious way.

However, the same cannot be said about the power requirements of the many different IP cameras, video servers, and other devices that an electronic security company may need to install to meet a client's particular needs.

A quick review of the plug-in transformers contained in the Security Networking Institute's training equipment bag reveals the following: out of twelve power transformers, no three are of the same voltage. Networking manufacturers are using power supplies ranging from 5.0 to 12 VDC, with no consistency whatsoever.

This issue is of great concern for security installations for a number of reasons. If using plug-in transformers, accessible non-switched AC power outlets must be located to power an IP-addressed camera, for example. Consideration has to be given to somehow protecting the plug-in transformer (also known as a "wall wart") from being unplugged. The "wall wart" must also be located somewhat near the camera, as the length of connection cables provided by the vendors is generally fifteen feet or less. Building electrical and fire code problems can arise if a plug-in transformer and its associated cabling are installed within a plenum ceiling. And even if a suitable, secure, and code-acceptable wall outlet is found, any power failure, brownout, or surge can either disable, destroy, or "lock up" the software in the device.

Everybody wants some . . .

The very devices that provide the tremendous flexibility of networking installations are just as susceptible to power problems. Routers, switches, hubs, and cable/DSL adapters all have their own power requirements, and can suffer the same types of problems listed above if there is a glitch in the power grid. What this means is that video or alarm signals and device communications can be lost due to power problems along the signal path.

Enterprise power concerns

Power problems can dramatically damage enterprise network functions. A sudden power outage can corrupt database files, ruin hard drives, and stop network functions. To counter this threat, an Uninterruptible Power Supply (UPS) is often found in network equipment rooms, to provide a smooth transition to battery back-up power in the event of a utility service problem. As the growth of networking electronic security applications continues, installing companies will become increasingly involved in power protection and UPS selection, particularly if the security dealer is installing a new or parallel network, providing switches, routers and other powered equipment to transport security information. Knowledge of UPSs and their functions is critical to the longevity of network security equipment.

What does a UPS do?

Before detailing the different types of UPS, it's important to understand the two basic functions that they are built to provide. First, a UPS protects connected equipment from power line surges, spikes, and distortion, which can damage sensitive computers and any software that's running when the power glitch occurs. The second UPS function is to provide backup power from its batteries if the utility power has failed completely.

Types of UPS

The UPS industry is well established and has developed an interesting and truly evasive marketing strategy. When visiting a computer store, there are many devices providing various levels of power protection and backup, at widely varying prices— and each product is labeled a "UPS." There are actually three distinct types of power products that fall under the general UPS category. As will be explained, each type of UPS provides some level of power protection and emergency backup, with differing grades of protection and sophistication.

Standby backup offline UPS

The "economy" model of UPS is generally termed a "standby backup offline" (SBO) device. These products provide direct connection between utility power and the "protected" computer equipment, with the same level of surge protection found in an inexpensive plug-in surge protector six-pack extension cord. If power fails, the UPS will switch on the battery backup, which is connected through an inverter that converts the DC power from the battery into 60-hertz AC. This emergency power is generally available for only a few minutes, which is enough time to shut down running programs and computers in an orderly manner. (Of course, this assumes that a quick-thinking and acting person is on hand at the time of the power failure.)

This least expensive relative in the UPS family has its flaws. The switchover time from utility power to battery backup (measured in milliseconds) can be long enough to "lock up" running computers and software programs. SBOs provide little or no assistance during a "brownout" condition, where utility voltage has dropped for longer than a few seconds.

Computers and network devices work best when they are receiving a true or pure sinewave form of 60 hertz AC power, as is generally supplied by utility services. When an SBO provides power from its battery backup, the waveform is often distorted into a square, modified square, or quasi-sinewave. This impure form of power can also cause problems with network computers and software.

Line interactive UPS

An upgrade to SBO technology is the "line interactive," or LI, UPS. These devices have all of the characteristics of an SBO, with the addition of a power line monitoring function, which can add power from the battery pack to make up for "brownout" voltage deficiencies. This voltage regulation feature is important, and worth the increased cost.

An LI UPS has a potential problem that should be brought to light. If an LI UPS expends its battery backup power to bring voltage levels up during a brownout and then utility power subsequently stops completely, the UPS may not be able to provide enough backup time for an orderly computer network shutdown.

On-line UPS

The top-shelf UPS is the on-line type, which provides a complete electrical firewall between any connected devices and utility power service. Incoming AC power is conditioned and provided in a three-step process.

First, the AC current is converted into DC and filtered through capacitors, which remove transients, harmonic distortion, and other unwanted elements. The backup batteries are connected at this first stage, so that when their power is drawn, it is filtered and conditioned by the on-line UPS before reaching the connected local devices.

The second stage provides voltage regulation and a second set of capacitors, which can store power and help to sustain voltage output during brownout conditions.

In the final step, the DC is converted into clean sinewave 60 hertz AC power, which is supplied to the connected devices.

Because of this three-step design, an on-line UPS provides transition to backup power with no start-up or delay time. Computers and software are fully protected from utility power anomalies because of the electrical isolation provided.

Battery backup

An inexpensive UPS has a built-in battery and will provide backup power for a few minutes. Higher grades of UPS can have additional battery packs added to them, increasing the potential standby power time.

Intelligent UPS

Just as is occurring within the electronic security industry, the UPS manufacturers are providing upgraded devices that include Ethernet connections and accompanying software programs. Connecting a suitable UPS to the network allows authorized users to monitor UPS functions from a PC and reset or change the device's parameters from the LAN or over the Internet. The UPS can be programmed to automatically shut down connected PCs and to apply file names and save open files prior to turning off active computers.

UPS capacity requirements

The capacity of a UPS is dictated by the amount of current being drawn by the connected computers and peripherals. UPS manufacturers typically provide "calculators" in their literature and/or on their web sites to assist in determining which particular UPS is needed for an application.

UPS application considerations

For electronic security installations, powered devices should always be connected to a UPS, if available. In the case where the installation includes new network transmission equipment such as switches and routers, the installing company may need to supply a separate UPS for backup power.

Remote control and reset

If there are problems with the utility power source, routers, switches, and even network cameras and video servers may "lock up" or shut down. The forward thinking security installation company will program its devices for remote access over the Internet, which may allow the resetting of devices remotely, without requiring an on-site visit by a service technician.

Network powering option

As the use of Ethernet networks expands, one of the exciting new technologies now available is called VoIP (Voice over Internet Protocol). VoIP replaces standard telephone communications, converting analog telephone calls into data packets and transmitting them over the Internet. As is detailed in Chapter 31, VoIP's attraction is greatly reduced costs for telephone communications.

When implementing VoIP, one of the issues is the matter of how the telephone instrument is to be powered. With a standard telephone connected to a POTS (Plain Old Telephone Service) line, 48-volt DC power is provided on the two-wire phone line from the telco central office to power the connected telephone instrument. Because this power is backed up from the central office, standard telephones will usually still work even during a local power failure.

When installing VoIP telephones, there is no power available from a typical Ethernet connection, so a power transformer or "wall wart" must be plugged in near the telephone. This situation causes two potential problems—where to plug in the transformer/power supply and the potential for phone failure during a power outage. Similar issues also can arise when installing Wi-Fi access points and other network devices.

Power over Ethernet ("PoE")

Recognizing this problem, the IEEE developed the 802.3af standard, which specifies how DC current can be coupled to pairs within the Ethernet UTP cable, providing voltage to power VoIP telephones. This standard is commonly termed Power over Ethernet, or "PoE." PoE is specified at 48 volts DC, with current limited to 350 milliamps.

How PoE works

Power is supplied to the same pairs that are used for transmission and receiving in 10/100 Ethernet.

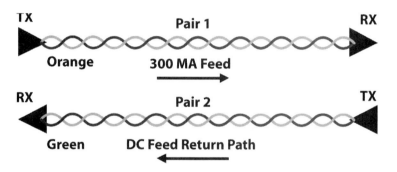

Figure 20-1   Power over Ethernet.

Devices capable of using the power, such as a VoIP telephone set, are termed "Powered Devices." The powered device has circuitry which taps the voltage from the Ethernet connection and supplies it to the device itself.

Although the unused two pairs of the four-pair UTP cable may also be used to transmit power, most manufacturers and end users are implementing the "same pairs" technology, leaving the unused spares for future Gigabit Ethernet upgrade potential.

PoE power suppliers

There are two methods of injecting power into the UTP Ethernet pairs. The first is called "Endpoint" power, with the power injection being included in the Ethernet switch itself.

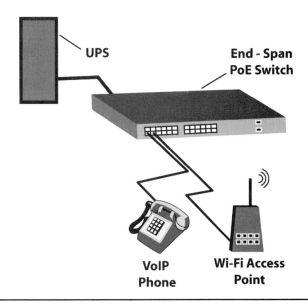

Figure 20-2 End-span powered switch.

This method is typically used if a new switch is installed.

For existing Ethernet switches, "Midspan" power injection can be added. This is a device that is separate from the switch and generally located directly under or over the switch in a rack-mount arrangement. Short UTP jumpers are connected from the ports on the switch to ports on the midspan power source. A second set of ports on the power source provide PoE-enabled outputs, which are connected to the structured cabling wall outlets.

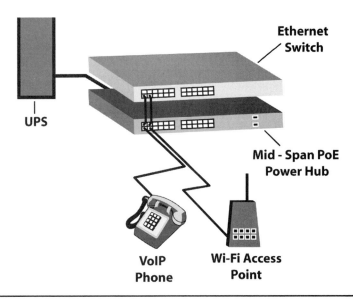

Figure 20-3 Mid-span powered switch.

Intelligent discovery

PoE systems cannot indiscriminately shoot DC voltage to all connected ports, which would cause damage to devices that don't require PoE power. To electronically sense PoE-powered devices, the powering equipment uses a simple technology called "Intelligent Discovery." When a remote Ethernet device is initially connected, a low-current voltage is transmitted onto the UTP. If the PoE circuitry detects a 25k resistor, full 48-volt DC will be supplied to that port. If the resistor is not detected, power to that port is not applied.

Advantages of PoE

The clear advantage to this technology is that the power for VoIP telephones and other devices can be supplied centrally from PoE-equipped Ethernet switches. When primary power to the PoE switch is connected to a suitable UPS, backup power is available for all connected devices. This eliminates the need for separate power supplies for connected devices.

> **SECURITY TECHNICIAN'S NOTE:** If planning to add PoE to an existing cabling plant, be cautious of the type of UTP cabling you are using. Some inferior grades of Cat 5e cable have smaller copper conductors and cannot provide the full current for the rated distance. Also, when planning to install PoE cameras or devices, make sure that the distance from the device to the PoE hub does not exceed 100 meters.

Growth of PoE

It is clear that this is a technology that has tremendous potential. Now the cabled Ethernet network, which can connect all manner of computers and communication equipment, can also supply power for many of those devices. This greatly reduces or eliminates power-cabling issues and provides a clean and potentially battery backed-up power source for network devices.

The largest manufacturer of Ethernet switches, Cisco, is a strong proponent of PoE, as the company is also a large presence in the VoIP market. As of the end of 2003 Cisco shipped over 60 million PoE-enabled Ethernet ports.

DC pickers

The availability of PoE will become a tremendous benefit for networked electronic security system installations. With one or two exceptions, electronic security equipment manufacturers have yet to build network cameras or video servers with PoE capability. Over time, this situation will surely improve, as major vendors such as Sony are already producing some PoE devices as of mid-2004.

Manufacturers in the computer peripheral equipment market are already producing adapters that will "tap" available PoE voltage from an Ethernet port and provide outputs to power non-PoE compliant equipment. These adapters plug into the PoE driven RJ-45 socket and provide either a "passive" or "regulated" voltage tap. Passive taps provide 48-volt output, while regulated taps will convert the

48-volt PoE power to either 12 or 24 volt DC. These PoE tapping modules are also called "DC pickers" or "Active Ethernet splitters."

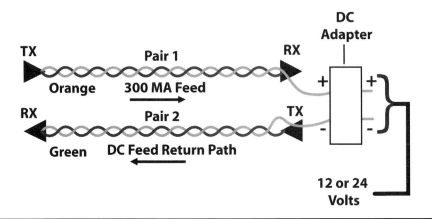

Figure 20-4 DC adapter.

As PoE grows in the marketplace, new and different power tapping devices will surely be developed by innovative manufacturers.

PoE and security systems

Although this is a new concept in powering electronic security devices, installation companies should embrace the availability of PoE. By utilizing this power source, cabling for device power and/or plug-in transformers can be reduced or eliminated, and security devices can be provided with backup power from centralized UPS sources.

Network Security

Introduction

The vast promise and potential of network applications for electronic security systems is tempered with a harsh reality—there is a dark horde of "hackers" and "crackers," inside and outside of the client's business, who can and will compromise, defeat, or disable network communications. A thorough knowledge of how hackers operate, and how they can be thwarted, is critical to the ongoing integrity of networked electronic security systems.

Why is this important?

The openness and innovation of computer networking and the Internet is its greatest strength, and also its greatest weakness. New products, web services, and software packages are introduced every day, all with the primary function of providing connectivity to other devices on existing networks and the Internet. All of this worldwide connectivity is a great thing, except when one of your systems is crippled by a "Denial of Service" (DoS) attack from a gang of hackers from anywhere on the globe!

When installing and maintaining network-connected security systems, the electronic security company has a doubly important responsibility to protect network communications from attacks. It is one thing for hackers to shut down an online retailer. It is of much greater concern if networked electronic security equipment is exposed to attack, which could both disable the security system and allow undetected physical entry into the client's premises.

Call the police

Estimates of the losses caused by network attacks and hacking vary widely, but the general consensus is that they cost individuals and businesses billions of dollars annually. The network security industry, which sells firewalls, anti-virus software, and other products, generates billions of dollars in sales as computer users strive to protect their networks from invasion.

Unfortunately, network attacks can happen at lightning speed, and hackers are well versed at disguising the computer they're using to initiate their nefarious attacks. These factors, along with the sometimes-nebulous amount of loss sustained by the victim, put law enforcement in an impotent position. Although the laws are stringent, and penalties severe when hackers are caught, unless it is a high profile incident the victim will have difficulty raising the interest of state or federal agencies. And if law enforcement gets involved, it can only provide prosecution after the damage is done.

The high chaparral

These issues place network administrators in a situation similar to the "Wild, Wild West" of American history. Imagine a cattle rancher in Wyoming, with thousands of cows, whose property is so large that he can't see the end of it from his house. The nearest town is ten miles away, and the sheriff's office is understaffed and of little help. If cattle rustlers are intent on taking the rancher's animals, our hero has no choice but to defend himself, his family and property using any means available.

Fencing the property is the first step to keeping the bad guys out, but checking for fence cuts and repairing them will become an everyday concern. And if the rustlers are determined, and the property is vast, some loss of cattle will likely occur. The rancher's goal is to minimize such losses.

The same goes for protecting networks from attacks or compromise. As the implementation of electronic security systems onto networks grows, more and more connected devices will be installed in remote locations, making them potentially harder to defend and easier to attack. Those with the responsibility for the integrity of electronic security signals connected to a network and/or the Internet must be aware of potential attack avenues, and construct "fences" to close off hacker opportunities. And, once these protective fences are in place, they must be regularly inspected to insure their protective functionality.

Who are the bad guys?

There are three types of individuals who may attack a network. The first is the "hacker," someone who enjoys manipulating computers and networks for fun, to incite mayhem, and possibly for profit. In the network community a debate rages over the concept of "good" hackers, sometimes called "white hats," who use their computer skills to find security holes and bring them to the attention of network administrators. The debate is about whether it's a good thing for anyone to attack a network, even if it's just to expose the holes. Is it right for a person to walk down a late night street, trying every door to see if it's locked? The white hats see their work as a public service, finding security flaws and bringing them to daylight so that they can be repaired.

That argument aside, this manual will use the term "hacker" in its negative connotation. Hackers are bad people, who want to exploit security weaknesses to the detriment of network integrity. Many hackers are young boys/men, who are experiencing the thrills of property damage, just as graffiti artists love to spray paint buildings. Some hackers are professional criminals, who use their skills to defraud individuals and corporations. Another factor in the lack of focus by law enforcement on computer crime is that many hackers are teenagers, and as such will receive minimal or no jail time in the event of a conviction.

The professional hacker is the greatest concern for the electronic security industry. Combine a far-flung security system, monitoring many buildings across a wide area, with a criminal hacker teamed with an adept burglary gang, and all of the pieces are in place for a major crime to be committed. The hacker attacks the network, "spoofing" or disabling security monitoring, and the burglars reap their rewards. Unless the security of the network is strong, this scenario is likely to occur.

Inside operators

As is often the case with other types of security compromises, the primary culprit in many network attacks is a disgruntled employee or "inside" person. Security guards and other personnel who become intimately familiar with a network security system are in the best position to commit violations to that system. With their continuous access to network computers and software, the nefarious inside person has the easiest opportunities to perform such attacks. As such, plans to protect a network should concentrate on preventing attacks from inside personnel as much or more so than from outside hackers. Keep in mind that these "inside" attackers may be employees of the electronic security firm, contracted guards, and/or employees of the protected company.

Why hackers might attack

We have already mentioned one reason why hackers might attack a networked security system, that being to enable a burglary or robbery to take place. While stealing physical property can benefit a criminal, the theft of intellectual property can have greater value to the bad guys, while damaging the protected client to a tremendous extent. Consider how easy it would be, given the proper access or a disabled security system, for a criminal to copy a company's customer list, product design specifications, or other mission-critical information, and carry it out of a protected premise on a single CD-ROM disk. Or a competitor to the targeted company commits a break-in with the sole purpose being to trash the network, causing the victim company's business to grind to a halt as it attempts to reconstruct the digital underpinnings of the enterprise.

The key issue that determines whether a particular system will be attacked by hackers is whether they can find and easily exploit holes in the target network's security. If the security of a network is robust, the casual hacker may well proceed to an easier target for his attacks. The same holds true for the criminal hacker . . . with so many businesses' networks vulnerable, why waste time on a "hardened" target when so many easier ones are available?

> **SECURITY TECHNICIAN'S NOTE:** While attacks on networks can be purely electronic, "social engineering" is also a major concern. Social engineers use deceptions and ruses, either over the telephone or in person, to get unsuspecting company personnel to provide their usernames, passwords, device addresses, and other information that the "bad guy" can then use to electronically attack a network. Even something as ordinary as the IP address of a video server can provide a social engineer with access to a system. Security technicians should carefully guard any and all information about a networked security installation, and such information should never be provided to any unknown person, regardless of who they claim they represent.

How hackers attack systems

When considering how to protect a network against hacker attacks, it makes good sense to have some idea how such attacks might occur. Here are some of the common types of hacker attacks.

Brute force attacks

One of the simplest forms of hacker assault is the "brute force" attack, which attempts to determine the username(s) and password(s) that allow entry into a computer or device on the victim network.

When a network camera is accessed, typically a username and password entry box appears, similar to the one shown in Figure 21-1.

Hackers use "brute force" programs that will repeatedly try different combinations of usernames and passwords until access is allowed past the sign-in screen.

Figure 21-1   Password Screen.

Some of these hacker programs are quite sophisticated, and will attempt various common passwords, such as peoples' names, until they are successful. This type of attack is also called "password cracking."

While the brute force password attack is probably the easiest to perform, if successful it can be devastating. There will be no warning from anti-virus software or the network, as the system has been compromised by using a valid username/password. Such an attack can also cast blame on the (possibly) innocent person whose password was cracked.

This type of attack can be thwarted by changing passwords regularly, informing users not to allow others to use their passwords, and not allowing passwords to be easily accessible to others, i.e. written on a Post-It note stuck on the side of the system's monitor. Password failure lockout is a software feature that locks out access to a system, for a set period of time, after a certain number of failed attempts to input a correct set of username and password. Unfortunately, some manufacturers of network electronic security products are not including this critical security component in their software programs at this time.

Computer viruses

A computer virus is a program or code string, written by a hacker, that will replicate itself onto other computers in a network, and may also try to reach other computers on the Internet by attaching itself to email attachments and executable programs that are shipped from one computer to another. Along with this replication ability, computer viruses will generally perform some system disruption. This disruption may be relatively benign, such as popping a message up on a web site or computer screen, or can be as severe as crashing a network server or erasing a hard drive.

There are three common ways that computer viruses enter a target computer. The first is through the downloading of virus-infected games and digital images from the Internet or bulletin boards. The second is email, where images that have a virus included may be attached or the email itself may include an infected executable macro file. The third way that viruses are spread is by the use of floppy disks. If a disk is used in one computer and then another, a virus can be picked up from the first machine and installed into the next.

Viruses can be readily downloaded from the Internet, and modified by hackers for their individual use. (Of course the hacker needs to be wary that the virus he or she has downloaded isn't itself infected.) Below are descriptions of a few of the most common types of computer viruses and their methods of attack.

Trojan horses

A "Trojan Horse" is a hacker's code string that has been connected to another program, typically a game or digital photo. Once downloaded by the user, the code will burrow itself into the target computer's operating system software, and may be nearly impossible to remove without reformatting the hard drive and reinstalling all operating and application software. Once in place, Trojan Horse viruses can create a host of problems. One common type of Trojan Horse will record all keystrokes, passwords, bank account numbers, and other critical information, and either deliver the information to the hacker when he accesses the victim computer, or transmit the information over the Internet to the hacker, all with no indication to the victim. "Back Orifice," a virus that attaches itself to Microsoft Office software, allows the hacker complete access and control of the victim computer from any Internet connection. These programs are powerful hacker tools, as hackers can take control of multiple computers, using one to control another, so that their true identity or location can be impossible to determine.

Logic bombs

This hacker attack is a set of software, which has infiltrated a computer and attached itself to an "executable" program, usually one of the many, many programs within the Windows operating system. Often hackers will use a permutation of a common Windows subfile name, such as "windll.exe," except the hacker code is called "wind1l.exe" or "windLl.exe." A cursory inspection of the Windows file register may not catch such subtle name differences. The logic bomb is activated when some other event occurs within the computer or network, or triggered remotely if the computer is remotely accessible.

For example, an employee installs a logic bomb in a computer or network that will erase the hard drive of a computer or server in the event that the employee is terminated. It is conceivable that a logic bomb in a network security system could be programmed to "freeze" the network when a certain access control card is presented to a reader, disabling monitoring and security functions.

The key to preventing logic bomb attacks is to deny hackers access to the network and its software. Such techniques will be detailed in a following section.

Worms

A "worm" is another virus variant, which will replicate itself by attaching its code to multiple executable programs within a single computer, and will seek specific security holes in other computers, where it will again restart the reproduction process. Worms are designed by hackers to attack specific software security flaws, usually those found in Microsoft products. Programs such as Outlook have been victimized in the past by well-documented worm viruses. Microsoft programs are typically targeted by hackers and virus developers, as their pre-eminence in the market means that a successful virus can infect the most systems.

Denial of service attacks

Another type of hacker attack requires no knowledge, embedded software code, or purloined network access by the hackers. All that is needed is the Internet IP address of the victim network, and a "Denial of Service" (DoS) attack can be launched.

The concept of a denial of service attack is very simple. The hacker uses a group of "robot" computers, which he's previously hacked into and taken command of, to send massive amounts of "ping" requests to the victim's network simultaneously. These robots are selected by the hacker for their broadband connection, and their relative ease of being hacked themselves. As the ping requests require a response, the victim network is so busy replying to the ping messages that it cannot perform any other communications. Hackers will modify the typical ping request so that the data streams sent are huge in bit size. If enough robot computers are sending enough elongated ping requests, the victim network's Internet connection will be filled with these spurious requests, not allowing room for other valid communications.

If a DoS attack is launched and sustained, the victim network's administrator will contact its ISP and ask the ISP to filter the ping request barrage at the router. Although this sounds simple, it can take hours to reach the right person at the ISP and convince him or her of the severity of the problem.

It can be nearly impossible to locate, much less prosecute the perpetrator of a DoS attack. The hacker will not use his or her own computer to bombard the victim, instead using multiple surrogates. While the sources of the ping barrage can be found from their IP address in the pinging packets, most likely that computer's owner/operator has absolutely no idea that his or her machine is doing anything unusual or illegal. Another legal problem, which applies to all hacker attacks, is that the Internet knows no international borders. For example, it can be extremely difficult for a US company to interest the Sri Lanka law enforcement authorities in arresting and prosecuting one of its citizens who has been accused of orchestrating a hacker attack on the US firm's network.

Spoofing

By programming a computer to pretend to be another, the hacker can "spoof" a victim's network into allowing him access. IP addresses are easily changed, and even MAC addresses can be faked over the Internet. A network may well allow communications and access to a "spoofing" computer if it presents the correct identification information when challenged.

Video spoofing

Although as of the writing of this guide it has only occurred in movies, the potential for spoofing of network security video signals should be a critical concern for our industry. Imagine a scenario where the monitoring location is in Chicago, and one of the cameras being monitored is outside Tucson, Arizona, watching a rarely used vehicular security gate at a high security plant. The analog camera is connected to an IP video server manufactured by brand "ABC." An insider in the Chicago monitoring station provides the IP address of the Tucson server to a confederate, who then purchases an identical video server from ABC. The criminal uses a video camera to record a few hours of the security gate scene, showing no activity. When the crime is to be committed, the bad guys disconnect the existing server by cutting the communication line to the Internet, and connect to the Internet the newly purchased video server, programmed to spoof the disabled unit. The previously recorded video signal is fed into the rogue server with a VCR. While the monitoring station sees the typical "no activity" scene, the criminals are committing the crime.

Although this scenario was seen in movies such as "Ocean's 11," it is a very real possibility, and one with a potentially disastrous impact on the electronic security industry. It doesn't take long to imagine how such a crime could be committed, particularly if an insider in the monitoring station is assisting in the process. The networking capability of the Internet means that the rogue server in the scenario above doesn't have to be located in Tucson; in fact, it might be in a room in Chicago right next to the monitoring station, being fed the appropriate video signals straight off of the system's NVR.

The potential for video spoofing needs to be recognized by equipment manufacturers, and quickly remedied. Identification and validation of video servers and their signals needs to be more than just having the correct IP address; at a minimum the MAC address of the server should be electronically confirmed on a regular schedule. Further, some level of encrypted machine-to-machine password should be periodically transmitted, ensuring that the device feeding the video signals is indeed the proper one.

A "hole" lot of trouble

Security holes can provide a ready path of hackers and criminals, who can exploit such an opportunity in many ways, some of which are detailed above. With all of these potential problems, the question arises as to whether electronic security systems should be connected to IP networks in any fashion.

It's important to remember that there is nothing completely secure in our world, regardless of the level of electronic or security guard protection that has been deployed to protect it. Banks get robbed, presidents have been assassinated, and jewelry stores have been burglarized. In fact, the first "hacking" attack on a security system may have occurred when an Apple II computer was used to spoof the coded multiplex signal coming from a central station-monitored alarm transponder in a jewelry store. This incident occurred in the late 80s/early 90s, with a well-respected national alarm company as the victim, along with the store itself.

Continuing on this theme, some would think that in a truly secure world, 24-hour armed guards would protect each and every property. When considering this concept, think about this: what happens if the guard himself decides to rob the property he's protecting? Even 24/7 surveillance cannot stop every crime from being committed, although it will deter most. And deterrence and vigilance is what electronic security companies provide.

Consider that the real function of access control, CCTV, and alarm systems is the replacement of guards, and their related costs. If guards are located at each entry door of a business, they can personally check in each employee or visitor as they enter and leave. The guard will watch for unusual activity, and call the police in the event of trouble. But guards are expensive to pay and manage so electronic security devices and systems have replaced them.

Networking electronic security systems takes this guard replacement economy to the next level, allowing far-flung locations to be managed and monitored from centralized locations. Networking security reduces security costs by using the Internet pathways that homes and businesses are already paying for to transmit alarms, video, and access control information.

The good news about potential hacker attacks is that an industry has sprung up to defend networks against such assaults. These vendors provide hardware and software products to defend computers, and to clean out viruses if they should appear. Many businesses and personal computers in homes are using these defensive software programs.

The other good news is that virtually all IT managers are very aware of potential hacker attacks, and are already providing high levels of security for their networks. If you are connecting IP-addressed security equipment to the client's network, sharing bandwidth and Internet connections, the protection provided by the IT department's defensive efforts will also provide some protection for your system's communications. Just be aware that sometimes these protections put barriers in between networked security devices and their data's destination.

Protecting your network

In the event that the electronic security devices are connected onto their own separate network, security contractors should be prepared to take an active role in planning, installing, implementing and monitoring the integrity of the network. Below are a number of suggestions for increasing the level of security of an electronic security network. These suggestions are by no means to be considered the final words on this topic, and security contractors must keep abreast of current threat levels and protective measures.

While considering these security measures, remember that the computer network within your own company may be vulnerable to sabotage, from within or without. Taking some precautions now may well save much trouble at a later date.

Network protection planning

Network protection planning should include:

1. Physical security of equipment
2. Security of communication lines
3. Defense against outside attack
4. Defense against inside attack
5. Protection of video data
6. Regular backup of system data
7. System restoral capability

Let's look at each of these concepts in detail.

Physical security-equipment

Easy access to routers, servers, DVRs, and related network security components is an open invitation to attackers. The simple disconnection of a network cable or camera can disable critical image viewing, while the removal of a hard drive can compromise video evidence. Equipment should be installed in locked cabinets, preferably within an access-controlled room or environment.

When planning an installation, it is important to consider back-up power requirements for network security equipment. Remember that a power outage may occur at any time, and could provide the same results as a DoS attack on the security network.

As most network devices are powered by plug-in transformers, security integrators should either provide a UPS with sufficient power to maintain the system's functionality, or use an existing UPS that is powering the client's enterprise network equipment. A related security topic is protection of plug-in transformer connections to AC power. A network camera's transformer, plugged into a standard wall socket, can be quickly disabled by simply unplugging the transformer. Using cameras that are powered using Power over Ethernet (PoE) is one option that provides higher security against such attacks. Many IP cameras and video servers can also be powered by separate power supplies providing twelve or twenty-four volt output.

Security of communications lines

The size and complexity of a network security system can create many opportunities for hackers or intruders to compromise the system. All power and communications cables should be installed to minimize the potential for easy disconnection. If possible, cables should be installed in conduit or raceways. Using protective conduits will also minimize the possibility of the bad guys "reading" the magnetic field surrounding a copper communications cable.

It is tempting to use existing "Cat 5" wall jacks to connect cameras, transmitters, and video servers onto the network. But as easy as it is to plug the camera into a convenient, otherwise unused wall jack, it is just that easy to disconnect. And remember that Ethernet devices are not "supervised." In other words these communications paths are not monitored for integrity. If an Ethernet device is disconnected from the network, no alarm will be activated. The only indication of a problem will be the inability to get images from a camera at the monitoring station, which might not be discovered for some time. Network cameras should be installed with UTP cable run to the camera, preferably hidden in the wall, with an RJ-45 male jack installed on the cable end for connection to the camera.

Defense against outside attack

Protecting the security network from outside attack consists of two basic ingredients, those being reducing or eliminating outside access and periodically scanning computers for viruses.

Outside access into a computer network is allowed or disallowed by the programming of gateway routers and firewalls, which can either open or close TCP/IP ports to outside information requests. Although it may seem desirable to completely close off these ports, that would defeat one of the primary strengths of network security systems, that being the ability to access and manipulate systems over the Internet or WAN. Sophisticated routers can be programmed to allow certain types of traffic, with some allowing such communications only during certain times, or on certain days. Even inexpensive Wi-Fi routers have the capability of being remotely programmed over the Internet, allowing a security company to change router communications settings from its office without requiring a site visit.

To perform a basic test of the vulnerability of a security network system to an Internet-launched attack, visit www.grc.com. This web site provides a testing sequence that can warn a system operator of potential security holes that may exist, such as common TCP/IP ports that have been inadvertently opened to outside communications.

Firewall and anti-virus hardware and software play a critical role in stopping unwanted intrusions into a network. The key to using these protective elements is to keep the software current by downloading updates regularly, setting the software to block unwanted communications, and regularly executing a virus scan of any network computers.

The good news about potential virus attacks is that most of them are currently being transmitted via email attachments. Devices such as network cameras and video servers are not equipped to accept email messages; they have no email address. This protects these field devices from the most common virus attacks. Virus protection needs to be addressed for full-blown computers on the network, particularly those used for system monitoring and control.

Defense against inside attack

"Inside" people, such as security guards, IT people, and employees of electronic security companies, present the largest danger of system compromise. A guard sitting at a security console on the midnight shift may have hours of unsupervised time to attempt to alter the network. How can such attacks be prevented?

There are a number of simple and very effective methods to reducing the opportunity for mischief by inside personnel. The first is physical, and needs to be emphasized: ***Remove or disable all floppy, CD and DVD drives from computers in the monitoring station(s)!***

Eliminating removable media drives makes it nearly impossible for an inside criminal to bring a floppy or CD with a virus or system-compromising program into the monitoring area and install it onto the system. This security policy also removes the potential for an inside person to copy important information, such as password lists or access control parameters, onto a disk and carrying it out of the building.

If it is difficult or impossible to physically remove the drives, software programs are available that will disable media drives and USB ports. These programs allow users with the correct username and password to reactivate disabled drives when needed.

To allow for system updates and backups, a plug-in CD, floppy, or DVD drive can be temporarily connected to the monitoring computer and removed when the task is completed. Or possibly, these tasks can be performed over the network.

"Thumb drives" or "flash drives," which are small multi-megabit storage devices that plug into a USB port, are just as dangerous as CD or floppy drives. Use of such devices should be banned by security management.

Password protection

Often passwords and usernames are not changed from their factory default when a device is installed. When the username is "admin," and the password is "password," it's a simple matter for hackers to enter a system. All usernames and passwords should be changed from their default values.

Passwords should be complex, including capital and lower case letters, numbers, and symbols. Such complexity makes the success of a brute force password attack much harder to achieve. Passwords should not be a person's name, license plate, or other easily recognizable numbers or words. Passwords and usernames should be as many characters as is allowed.

Changing passwords and usernames on a regular or irregular schedule is another way to thwart hackers. Password files should be regularly checked so that any users who've been dismissed or changed positions have been deleted.

Eliminate or reduce Internet access for security personnel

Firewall software can be programmed to limit or eliminate the access of security personnel to the Internet. If there is no reason for a particular person or computer to be able to reach the Internet, why allow it? Stopping Internet access can keep personnel from reaching hacker sites and potentially downloading viruses into the network. An insider, using a security computer that allows free Internet access, could also download a pre-made virus or compromise program from an FTP website directly into the security computer without needing a removable drive. Disabling Internet access for security personnel doesn't mean that authorized users cannot reach the system from the outside.

Keystroke logging

There are a number of hardware and software devices that will provide "keystroke logging," recording every action performed by a user on a computer. Software programs can be set to transmit copies of all emails to a remote location as well as logging all activities. Inexpensive hardware devices can be plugged into the back of a computer tower and will record all keystrokes performed on a particular computer. After retrieval, the hardware device can be interrogated for all information or selectively searched for specific keystroke sequences such as "hacker software" or "DoS attack software."

Video monitoring

Recorded video cameras, preferably concealed, should be installed so as to record the actions and movements of security personnel as they interact with monitoring computers and equipment.

Aggressive spoofing

It is very important to keep a high security profile with inside personnel. Inform them that internal security is very important and that to protect the security system, anti-virus software, keystroke logging and other measures are being used. Many companies have reduced internal hacking incidents by announcing to their personnel the implementation of anti-hacking measures. If the insider feels he might get caught, he's less likely to commit the crime.

Employee screening

To prevent inside attacks on a network, electronic security firms and guard companies should carefully review their pre-employment investigative procedures. Drug testing, psychological profiling, and careful criminal record checking should be included when determining a candidate's viability for a security-sensitive position. Once a prospective employee has passed pre-employment screening, many companies do not perform follow-up periodic testing which might possibly expose an employee who has developed an addiction problem during his or her employment period. Any such testing must conform to applicable state and federal laws regarding employees' rights.

Protection of video data

A DVR located in the monitoring area is an open invitation for compromise or crime. An armed robbery gang can enter a building, perform a robbery, stop at the security desk on the way out, grab the DVR, and the video evidence of their crime is gone. DVRs in the monitoring area can be compromised by guards ("I didn't know that I unplugged it"), alarm technicians, or anyone with access to the area.

Ideally, digital video recording should be performed on equipment located away from the monitoring station, such as within an access-controlled computer room. Some products provide for video recording at remote locations connected to the network, so the video storage can be in New Jersey, with the monitoring station in Pittsburgh, and the video servers in Los Angeles. Sophisticated systems also can provide multiple storage streams so that critical video information can be stored in multiple locations simultaneously.

Regular backup of system data

A regular schedule of "backing up" system information is critical to the ability of electronic security companies to quickly restore a system that has been compromised. Such backup capability may also be necessary in the event of hard drive or computer failures. Backups should be performed regularly, and the disks or media should be stored off-site, perhaps in a fireproof safe at the security contractor's office.

Redundant systems/paths

In high-security applications, establishing a redundant monitoring station can make a critical difference if the primary monitoring station is attacked, or suffers a catastrophic power failure or other negative act of God. The power of the Internet, with its multiple transmission paths, provides great flexibility in planning for potential disasters. A properly programmed laptop computer could be connected to the Internet from another location and quickly receive video and take control of a security network when the main monitoring station has been disabled.

Network security is a process, not a goal

While implementation of the above security recommendations will reduce the potential of inside and outside attacks, securing a network should be a constant concern. System additions, network changes, and personnel changes all will occur with great regularity, and each such change can open potential security holes that can be exploited. Security measures should be regularly reviewed, tested and modified to protect the network against the latest threats. Anti-virus programs should be regularly run on monitoring computers, and system back-ups should be performed regularly

DSL Adapter Guided Tour

"Adapters" provide communication access to the WAN or Internet for local computers, networks, and devices. This guided tour of a DSL adapter will highlight the programming options that are relevant to electronic security system installations.

DSL adapter functions

DSL, or "Digital Subscriber Line," is a telephony-based service that provides broadband Internet access to home and business subscribers. Frequency bands are used for uplink and downlink Internet communications, while analog voice calls can be transmitted at the same time.

Common DSL adapters provide communications functions between the ISP's network and the subscriber's computer or LAN network. Typical DSL adapters provide a relatively simple wiring connection: two RJ-45 female jacks, one for the LAN and one for the telephone line connection, and a power input, usually fed from a plug-in transformer.

Some DSL adapters also provide firewall features, protecting LAN computers and devices from spurious or potentially harmful communications. These features must be addressed by the security installation technician if devices such as IP network cameras are to be viewed over the Internet.

The DSL adapter used for this tour is a standard small office/home office device that receives a single DHCP address from the ISP. See Chapter 9 for more information about DHCP addressing. More sophisticated adapters that provide static IP addresses operate more like a router, and will be accessed and controlled in a different manner.

Accessing the DSL adapter

For typical DSL adapters, a PC or laptop must be directly connected to allow initial access to the programming options. Just as the DSL adapter shields local network computers from the Internet, a gateway router connected to a DSL adapter shields the DSL adapter from the LAN computers.

To access a DSL adapter's settings, a laptop or desktop PC must have its NIC set up for DHCP, and must be connected to the LAN port on the DSL adapter using a UTP network jumper (normally, a "crossover" cable isn't used). Remember to write down the original IP settings for the programming laptop or desktop PC so that they can be restored after programming of the DSL adapter is completed.

If the local IP address of the DSL adapter is known, it can also be accessed from any LAN-connected PC, using a web browser program and the adapter's IP address.

Once the computer is connected, the DSL adapter can be accessed by either clicking a shortcut icon on the desktop, starting the program from the "All Programs" selection in the "Start" menu, or inputting the adapter's IP address into a web browser program.

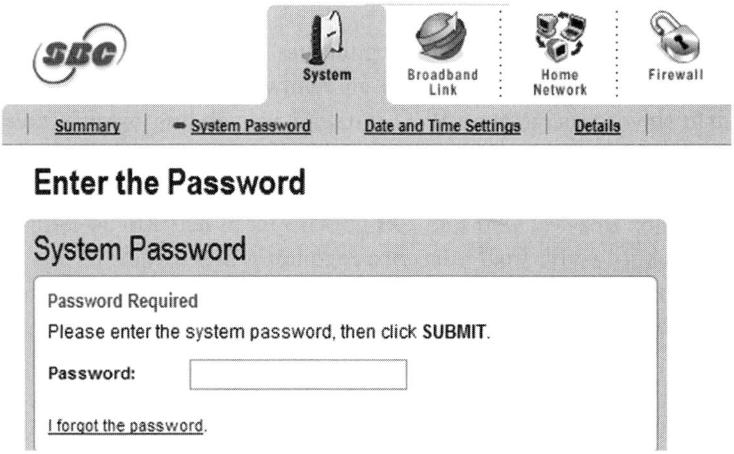

Figure 22-1   DSL sign-on screen.

Password access

In order to make any changes to the adapter's settings, generally a password or username/password combination is required to gain entry. This password was set by the user during the initial setup of the DSL connection. Notice that if the password has been lost, there's a link that, when clicked, will contact the ISP, which will start a process to either email the correct password or set up a new one.

Firewalls and hosted applications

Computers connected to the Internet must be protected against intrusion by hackers. Some cable and DSL adapters protect the computers (and LAN networks) attached to them by providing firewall functions, which basically stop any incoming traffic from unknown sources. This incoming traffic is distinct from email and web page information, which has been specifically requested from the Internet by one of the computers on the LAN.

There are instances when it is desirable to allow traffic into the network from the outside. Many users enjoy playing interactive video games, which require two-way communications between various players. Other two-way technologies include VoIP phone adapters, FTP web sites, and web page hosting.

Of particular interest to the electronic security technician is the connection of network-enabled DVRs, video servers, and IP cameras. Access from the outside must be allowed for remote viewing, programming, and control of these devices.

DSL and cable adapters provide two methods of allowing unsolicited traffic to enter into the network from the Internet. Those two methods are generally termed "Hosted Applications" and "DMZ," or "de-militarized zone."

Hosted applications

A local device (or program) that is to be allowed two-way communications through the firewall is called a "hosted application." In this case, the local device will act as a "host," providing programs or data to other computers that access the host, using the proper username and password.

To allow communications through the adapter's firewall, the technician must add the hosted application to the adapter's firewall table by programming the host name, protocol, and port number(s) that have been selected for hosting use. See Chapter 9 for more information about ports and port numbers.

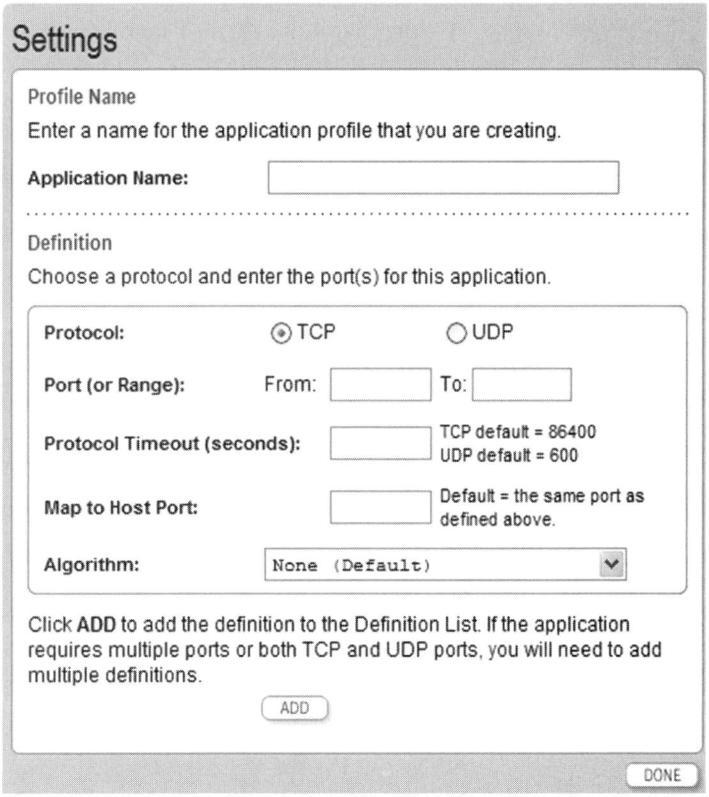

Figure 22-2 DSL edit applications screen.

After accessing the proper screen, a selection window similar to the above will be displayed. After entering the application (host) name of the IP camera or video server, the installer programs the protocol, which is usually TCP for most IP-addressed video devices. Some devices require more than one port for communications, so the adapter provides a "range" of port numbers. For most IP video devices, there is only one port required. In our example, port 85 will be used, so the entries for the "Port Range" fields are "85" and "85." Other default selections for "Protocol Timeout," "Map to Host Port," and "Algorithm" are generally left as is.

Once the proper information has been input, clicking the "add" button, followed by the "done" button, will add this hosted application to the adapter's firewall table.

Turn on the host

Once the "hosted application" has been input into the adapter's table, it must also be assigned to the proper computer or router to "turn on" the desired communication functions.

At this point, please recall that the DSL adapter may have been wired directly to a PC for programming purposes, as described above. This connection will be physically disconnected once the programming of the adapter is complete. The hosted IP camera is not going to be connected directly to the PC that is currently communicating with the adapter; it will most likely be connected to a gateway/router, along with the home or business computers. So it is not just a question of "turning on" the hosted application; it must be assigned to the proper network device, in this example the gateway/router.

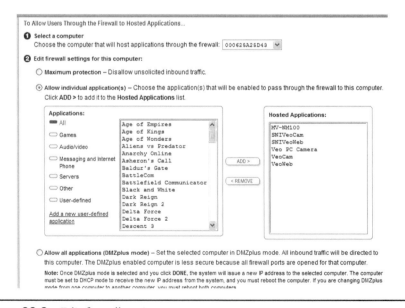

Figure 22-3 Edit firewall setting screen.

Select the computer

In Figure 22-3, the first step is "Select the Computer," with a pull-down box. When the box is opened, each computer or router that the DSL adapter has been or is currently connected to will be displayed. These computers or gateways will be identified by their MAC address or host name. The gateway that will be connected to the IP camera/video server must be selected by the technician at this point.

> **SECURITY TECHNICIAN'S NOTE:** Most DSL adapters don't allow the technician to input the MAC address of a gateway/router; the adapter must "see" the router itself. In an installation where a new gateway/router is being added, it must be plugged into the DSL adapter, powered up, and then disconnected before a direct PC-to-adapter connection is made for programming purposes. This will allow the DSL adapter to register the gateway/router as a viable network device.

Allow individual applications

The second step in programming hosted applications into the DSL adapter is to choose "Allow Individual Applications." In the box on the left, a large number of preset applications are already programmed into the adapter; scrolling down in this box will display the IP camera/video server application that was previously programmed. Highlighting the desired application(s) and clicking "add" will place the application in the "Hosted Applications" box. Clicking "Done" will complete the opening of the proper ports for the passing of communications to and from the IP camera through the DSL adapter's firewall.

> **SECURITY TECHNICIAN'S NOTE:** Both the DSL/cable adapter AND the gateway router must be properly programmed to allow two-way communications to a specific IP camera or video server. If one or the other isn't set up correctly, no or limited communication is possible. The technician can start at one device or the other, but both must be programmed.

More than one camera?

If additional devices need to be programmed to pass through the adapter, the steps listed above would be repeated. If, for example, two DVRs need to be accessible from the Internet, each would be individually programmed as a "hosted application." Each DVR would need to be addressed to a different port number, to allow each to be contacted individually.

DMZ mode

Notice the last selection on the screen illustrated in Figure 22-4. Sometimes it is desirable to allow all traffic from outside the LAN to reach a single network computer or device. Placing a computer or gateway/router into the DMZ allows all traffic from the Internet to reach the selected device without any firewall filtering. This setting is sometimes used for connections such as a PC set up as a public File Transfer Protocol (FTP) site. This setting may also be needed when installing a VoIP (Voice over Internet Protocol) adapter, to allow the adapter to be constantly controlled by the VoIP service's network server.

Review the work

Accessing the "Firewall Details" screen provides a review of what has been programmed into the adapter for hosting. The "Device" number is the MAC address of the gateway selected to host the applications; other information displayed is the host name, port number, and a "Public IP" address. This is the Internet IP address of the DSL adapter, which in this case is a DHCP address that can change at any time at the whim of the ISP network's DHCP server.

Getting back to normal

After confirming the DSL adapter's firewall settings, close out the programming access and the PC's Internet browser. Reset the programming computer's IP address back to its original settings, and reconnect the cabling between the adapter and gateway router.

Review of DSL adapter setting adjustments

Any communications that must pass through a DSL or cable adapter stemming from the Internet must be allowed to pass through the adapter's firewall. Individual devices must be programmed into the firewall table as "hosted applications," and communication enabled to the specific gateway router to which the hosted device is connected. While placing a downstream gateway/router into the DMZ will allow all applications to communicate, this can expose the selected device to hacking from the Internet.

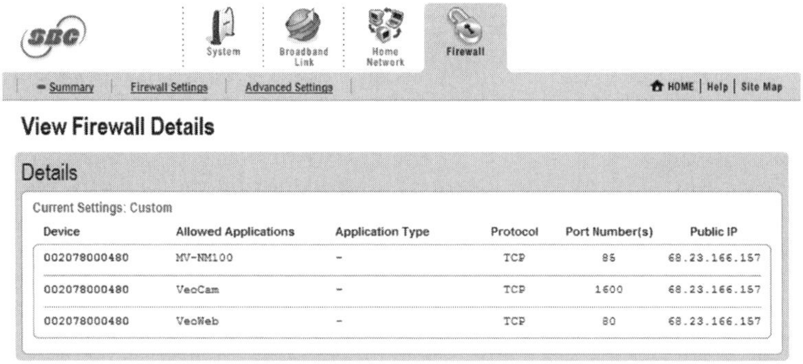

Figure 22-4 DSL adapter View Firewall Settings screen.

SECURITY TECHNICIAN'S NOTE: As is noted on other guided tours in this section, it is very important to either write down or print out the original settings of a DSL adapter before making any programming changes. If things go poorly, at least the original functionality of the system can be restored. It is important to test each connection or change as it is implemented to ensure communications and reduce or eliminate the need for troubleshooting.

DSL adapter review

DSL adapters must have their settings manipulated to allow "hosted applications," which opens the appropriate ports for remote users to access network devices connected to the DSL adapter. Different DSL providers will use various manufacturers' adapters, and security technicians should make themselves familiar with the specific types of adapters they may encounter in their local area.

Wi-Fi Router
Guided Tour

There are many programming options available in networking devices such as routers and switches. The following section takes us on a tour of a typical Wi-Fi router, such as might be used in a small business or office. Important options for security installations will be highlighted and described.

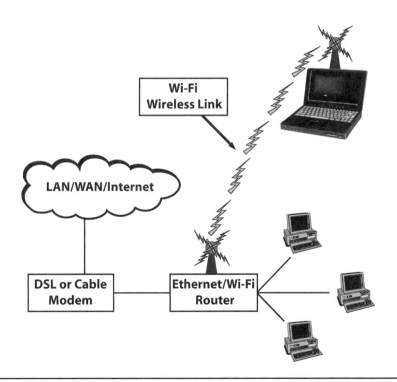

Figure 23-1 Wi-Fi router.

The router

Although commonly called a "router," the type of device described in this section is more properly termed an Ethernet/Wi-Fi gateway switch. The particular device described in this section is a Dlink #614+, which is a popular brand of these products. This device allows the connection of up to four wired Ethernet nodes (computers, IP cameras, etc.), plus Wi-Fi capable computers and devices, as well as an Internet connection, such as a DSL adapter or cable modem. This piece of equipment is typical of the type that would be installed in a small business or home to provide Wi-Fi coverage.

Accessing the device from a PC

Let's assume that the device is currently configured, connected, and functioning properly. To access the router's programming, open an Internet browser such as Internet Explorer, and type in the router's LAN IP address. If the address is correct, this window shown in Figure 23-2 will appear.

Figure 23-2   Password screen.

Typing in the correct username and password will allow the user access to all programming functions.

Wireless settings

The first stop in our router tour takes us to the "Wireless Settings" page, where most of the parameters for the Wi-Fi functions can be set. Let's take a look at Figure 23-3.

Turn on/turn off   The first selection from the top is whether the Wi-Fi network is "Enabled" or "Disabled." This feature can be used to quickly turn the wireless network on or off without having to completely re-input all settings. This can be useful in circumstances where a network operator may only want the Wi-Fi to be enabled on command.

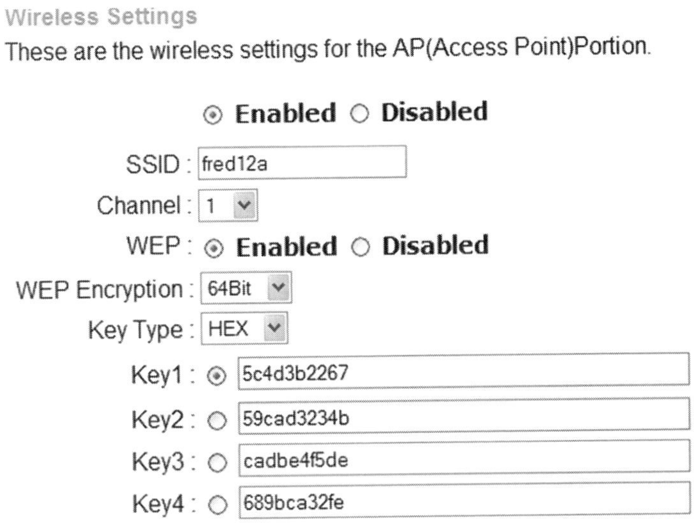

Wireless Settings

These are the wireless settings for the AP(Access Point)Portion.

⊙ **Enabled** ○ **Disabled**

SSID : fred12a

Channel : 1 ⌄

WEP : ⊙ **Enabled** ○ **Disabled**

WEP Encryption : 64Bit ⌄

Key Type : HEX ⌄

Key1 : ⊙ 5c4d3b2267

Key2 : ○ 59cad3234b

Key3 : ○ cadbe4f5de

Key4 : ○ 689bca32fe

Figure 23-3   Wireless settings.

SSID name

The next selection is the "SSID," which is the "name" for the particular Wi-Fi network, in this case "fred12a." This name is part of the association between the Wi-Fi router and the wireless computers or nodes that are connected. Wi-Fi routers can be programmed to broadcast this SSID name, telling Wi-Fi-equipped computers what name to use for connection. Most routers can be programmed to disable this broadcasting feature, which enhances network security. Computers (or Wi-Fi cameras) can either receive the SSID through the broadcast or have the SSID name programmed into them manually. SSID names should always be changed from the default, which is usually the manufacturer's name or the part number of the router.

WEP

Below the SSID name is a selection to enable or disable WEP, which stands for "Wired Equivalent Privacy." This is the encryption algorithm available in standard Wi-Fi products, which converts transmitted data packets into strings of unintelligible bits that can only be properly used if the receiving computer or router is programmed for the same WEP "key," which decrypts the data packets.

The WEP key being used will be common to all connected computers at a given time. The WEP key must be manually input into Wi-Fi computers and cameras before they can communicate with the router and other computers on the network.

WEP encryption and key type

WEP encryption can provide different levels of encryption. In the case of this router, the selections are 64-, 128-, and 256-bit encryption levels. Increasing the encryption level makes for longer data packets, and increases the difficulty of an outsider "cracking" the encryption code by sampling packets out of the air. Higher encryption levels also slow down data communications. WEP keys can be formulated in either ASCII or "Hex" (0-9, A-F) formats.

Below the "Encryption" and "Key Type" selections are four fields where the system operator can input unique character sets that will each constitute a WEP "key" that must be input into the Wi-Fi computers or devices that will connect with the WEP-enabled router. There are typically four different keys, which allows a system administrator to pre-set up to four different master encryption keys and distribute them to authorized users. When the WEP encryption is to be changed, the administrator can then inform such users to change the encryption setting on their computer from key #1 to key #3, for example. This eliminates having to re-enter multiple strings of potentially confusing WEP key codes.

SECURITY TECHNICIAN'S NOTE: Different products from different vendors use different methods of entering WEP encryption keys. In some cases, certain Wi-Fi network cameras will not allow the setting of their WEP codes in the same way as a particular access point or router. Also, remember that WEP encryption will slow down data transmissions, which will reduce frame rates from Wi-Fi cameras. It's a good practice to first program a Wi-Fi camera and router "in the clear," without WEP enabled in the router/access point or camera. After communications are established, then the related devices can be programmed for WEP functions.

Once the proper settings are selected/input, the settings are typically "saved" or "applied" to the device.

SECURITY TECHNICIAN'S NOTE: Remember to write down the settings, particularly if something has been changed. There are many options in these devices, and changing one or a combination may result in partial or complete non-functionality of the system. The device may need to be reset to the original settings to restore communications. This rule applies to all programming of network devices.

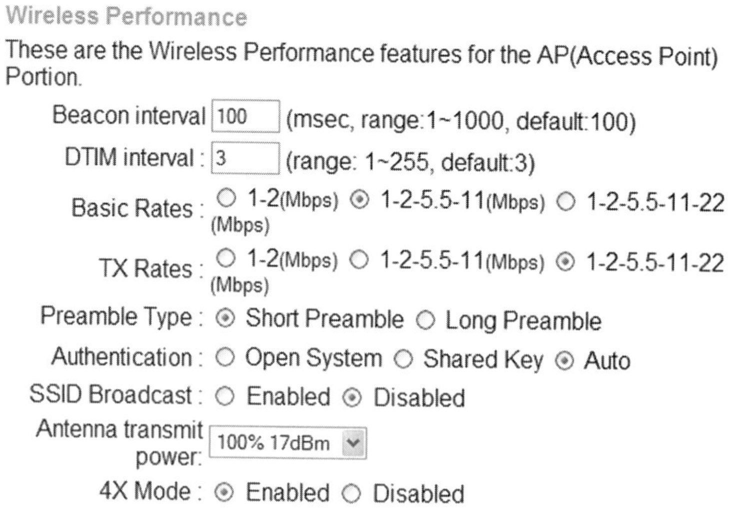

Wireless Performance

These are the Wireless Performance features for the AP(Access Point) Portion.

Beacon interval 100 (msec, range:1~1000, default:100)

DTIM interval : 3 (range: 1~255, default:3)

Basic Rates : ○ 1-2(Mbps) ⊙ 1-2-5.5-11(Mbps) ○ 1-2-5.5-11-22 (Mbps)

TX Rates : ○ 1-2(Mbps) ○ 1-2-5.5-11(Mbps) ⊙ 1-2-5.5-11-22 (Mbps)

Preamble Type : ⊙ Short Preamble ○ Long Preamble

Authentication : ○ Open System ○ Shared Key ⊙ Auto

SSID Broadcast : ○ Enabled ⊙ Disabled

Antenna transmit power: 100% 17dBm ▾

4X Mode : ⊙ Enabled ○ Disabled

Figure 23-4 Wireless Performance screen.

Disabling SSID transmission

A simple security measure for Wi-Fi networks is to disable the broadcast of the SSID network name. In this particular router, the ability to turn off the broadcast is tucked into a screen called "Wireless Performance," which includes other settings that can be manipulated, but shouldn't unless the system operator is well versed in Wi-Fi settings. Play if you must, but remember what the default settings were, so that you can reset the changes you made that didn't work.

Figure 23-5   WAN Settings screen.

WAN settings

The "WAN" settings are where the router is programmed for how it will be addressed by the Internet adapter, such as a DSL or cable modem connection. The two most typical selections will be either "Dynamic IP" or "Static IP." Low-cost DSL and cable connections use "Dynamic," which provides a temporary "leased" Internet IP address from the ISP for the router and network.

PPPoE stands for "Point-to-Point Protocol over Ethernet," and may possibly be needed for some DSL and cable modem connections. If PPPoE is selected, the router software will ask for a username and password, which it will then transmit upon request by the ISP's network.

SECURITY TECHNICIAN'S NOTE: How do you know if the client's ISP requires PPPoE? Here's how to find out. Open an Internet browser and type in a valid web address, such as www.slaytonsolutionsltd.com. If a password screen opens every time the Internet is accessed, the ISP is using PPPoE verification. If PPPoE is required, get the username and password from the ISP.

Host name Computers on a network can have common names as well as IP addresses. It's easier to remember "Dave's PC" than an IP address such as 192.168.2.15. Generally this setting can be left at the default.

MAC address The Media Access Control (MAC) address on this screen is the WAN MAC, providing a unique equipment identifying code to the network. This code should not be changed by the user, except as detailed below.

MAC cloning Connection of a wired or Wi-Fi router to a broadband Internet connection allows many users to receive Internet services from a single DSL, cable modem, or other ISP connection. Put in a Wi-Fi router and one broadband connection can cover a whole building, or a cluster of homes. This can potentially cut into the ISP's revenues, as one connection is serving many users.

While some ISPs don't care about connection sharing, some try to limit it in the following way. The ISP specifies that only single PCs may be connected to its adapters. When the DSL or cable modem is initially installed, the network will pick up the MAC address of that single computer and associate it with that particular connection. If another device, such as a Wi-Fi router, is connected to the adapter in front of the previously connected computer, the ISP network will refuse the connection, as the MAC of the router is different from the previously registered MAC of the PC.

Most common routers offer "MAC cloning" to defeat this constrictive technology. If selected, the router prompts for a new MAC address, which allows the operator to input the MAC of the previously connected PC. Now the ISP network believes it is communicating with the original PC, not the router. The People have the Power!

DNS addresses Domain Name Server (DNS) addresses are the servers that translate common term IP addresses, such as www.slaytonsolutionsltd.com, into their true IP numeric equivalents, in this case 205.158.155.152. These primary and secondary DNS addresses can be obtained from the ISP, or will be downloaded to the router if "Dynamic IP" is selected above.

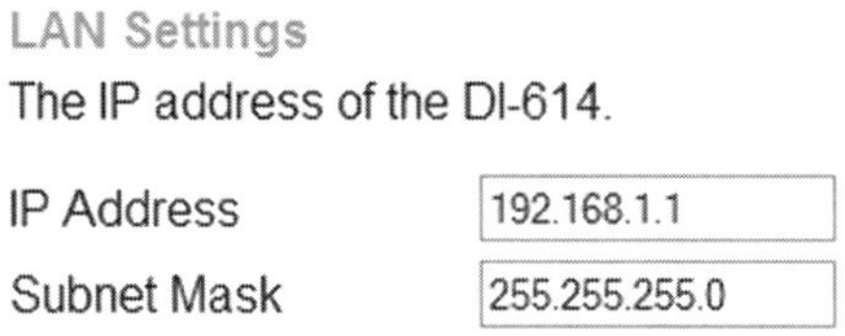

LAN Settings
The IP address of the DI-614.

IP Address	192.168.1.1
Subnet Mask	255.255.255.0

Figure 23-6 IP address field.

LAN settings

This screen shows the local area network ("LAN") address that the router will present to other computers, wired or Wi-Fi, within the local network. Generally, it's a good idea to leave this at its default value, which will usually be either 192.168.1.1 or 192.168.2.1. Inexpensive routers provide limited options for such features as Network Address Translation (NAT), and DHCP functions. (See the next section for more information on DHCP.) If the LAN address of the router is changed from its default "class C" network address, it may be impossible to set the NAT for cameras or other devices that must be viewed from outside the local network.

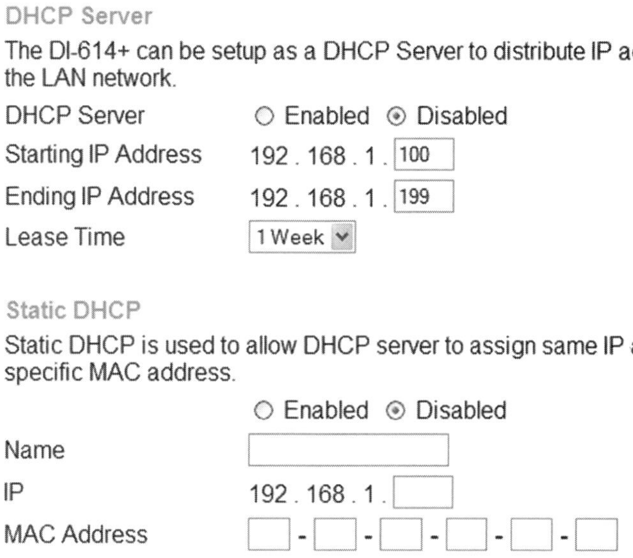

DHCP Server

The DI-614+ can be setup as a DHCP Server to distribute IP addresses to the LAN network.

DHCP Server   ○ Enabled ⊙ Disabled

Starting IP Address   192 . 168 . 1 . 100

Ending IP Address   192 . 168 . 1 . 199

Lease Time   1 Week

Static DHCP

Static DHCP is used to allow DHCP server to assign same IP address to specific MAC address.

                       ○ Enabled ⊙ Disabled

Name

IP   192 . 168 . 1 .

MAC Address   □ - □ - □ - □ - □ - □

Figure 23-7 DHCP Server screen.

DHCP settings

Even inexpensive routers have powerful network features. This screen provides the settings for Dynamic Host Configuration Protocol (DHCP), which instructs the router to provide Dynamic LAN IP addresses to computers and devices on the local network. This is a commonly used technology, as occasional network users don't have to be given a specific LAN IP address for input into their laptop. Visiting users just program their computer to accept a DHCP address from the router, and they're connected.

The first selection simply enables or disables the DHCP function. The "Starting" and "Ending" IP Address selections allow system operator control of just how many computers can be connected using DHCP at a single time. This can be used to provide another level of network security. If there are three wired

Ethernet computers and a maximum of two Wi-Fi laptops that may be connected at any one time, the operator can limit the DHCP addresses to a range of five. No other computers will be granted a DHCP address if all are currently in use, blocking the outside computers from accessing the network. The "Lease Time" is the time frame that a DHCP address is valid; if a computer needs to stay on the network longer, it must request and receive a renewed DHCP address from the router.

Static DHCP

Another level of security can be achieved by using "static DHCP." This setting tells the router to only issue DHCP addresses to computers with specific MAC addresses. A business can enter the MACs of the laptops of regular visiting users, such as upper management or salespeople, and they will be allowed onto the network, while non-registered laptops will be denied access. This is sometimes called MAC authentication.

> **AUTHOR'S NOTE:** The previous illustration shows two examples of how low-cost routers can be limited in terms of their LAN address options. Notice that the selections for "Starting/Ending" the DHCP field and "IP" under Static DHCP are all preset as "192.168.1. (fill in the blank)." In the previous "LAN Settings" screen, had the LAN IP been changed to something other than an address starting with "192.168.1.?," the DHCP options cannot be made to function.

Advanced settings

The previous router settings are those typically used when setting up a simple router/Wi-Fi system. Most routers have an "Advanced" setting section, which electronic security technicians may often need to manipulate to achieve desired access and network security features.

Customized Applications	Ext.Port		Protocol TCP	Protocol UDP	IP Address	Enable
SNIVeoWeb	81	To 81	☑	☐	192.168.1. 102	☑
SNIVeoCam	1601	To 1601	☑	☐	192.168.1. 102	☑
VeoPCCam	80	To 80	☑	☐	192.168.1. 110	☑
MV-NM100	85	To 85	☑	☐	192.168.1. 10	☑
	0	To 0	☐	☐	192.168.1. 0	☐

Figure 23-8   Port forwarding.

Port forwarding

Network Address Translation (NAT) provides the ability to access devices, such as cameras, from outside the network, such as from the Internet. This screen from a Linksys Wi-Fi router shows the necessary settings to achieve NAT for a particular LAN device.

To enable a camera to be accessed for remote viewing, "Port Forwarding" must be properly programmed. In the example above, the IP network camera named

"MV-NM100" has been set up in the router so that requests from outside the LAN addressed for port #85 will be forwarded to IP address 192.168.1.10. The "Ext. Port" is the port or range of ports that the camera will use. Typically, cameras will use TCP (Transport Control Protocol), as opposed to the User Datagram Protocol (UDP). Notice that the selections for the LAN IP addresses are limited to those starting with "192.168.1."

> **SECURITY TECHNICIAN'S NOTE:** Usually the setting information required for port forwarding can be determined by the particular camera or device being connected. Some experimentation may be necessary, as different cameras and routers will use different terminology to name identical features.

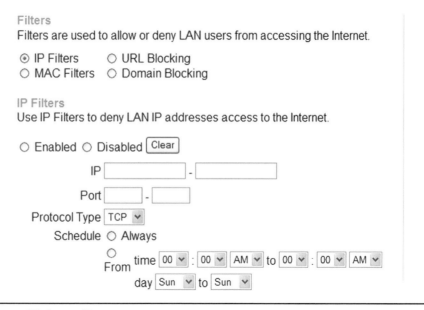

Figure 23-9   IP filters.

Filtering options

Many routers will provide the ability to enable or deny individual computers or devices access to the network, and/or access to the Internet. In Figure 23-9, "IP Filtering" allows the system operator to deny certain computers access to the Internet, either by a time/day setting or always. The same sort of Internet access or denial can also be set by the MAC addresses of network devices. "URL" and "Domain Name" Blocking provides a level of parental-type filtering by denying access to specific web sites or sites including non-desirable words in their address.

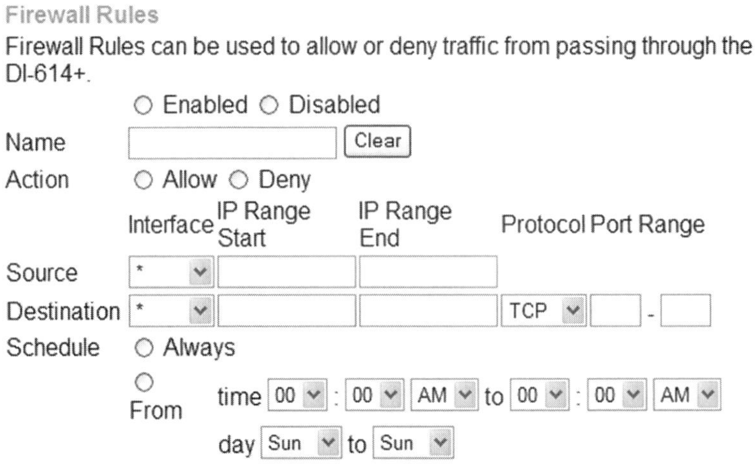

Firewall Rules

Firewall Rules can be used to allow or deny traffic from passing through the DI-614+.

○ Enabled ○ Disabled

Name [   ] [Clear]

Action ○ Allow ○ Deny

	Interface	IP Range Start	IP Range End	Protocol Port Range
Source	* ▾	[ ]	[ ]	
Destination	* ▾	[ ]	[ ]	TCP ▾ [ ] - [ ]

Schedule ○ Always

○ From time [00 ▾] : [00 ▾] [AM ▾] to [00 ▾] : [00 ▾] [AM ▾]

day [Sun ▾] to [Sun ▾]

Figure 23-10   Firewall Rules screen.

Firewall rules

This particular router provides settable firewalls between computers on the network. Perhaps a business owner doesn't want to allow his networked employees access to his computer, or maybe he only wants to allow such access during business hours. As with other communication options previously detailed, such firewall blocking can be set as "always" or on a time/date schedule.

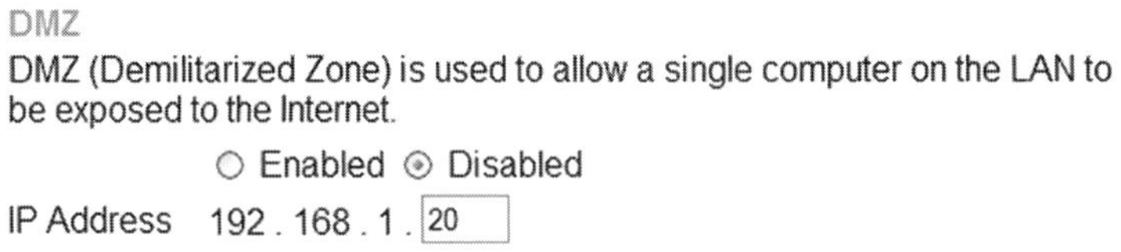

DMZ

DMZ (Demilitarized Zone) is used to allow a single computer on the LAN to be exposed to the Internet.

○ Enabled ⊙ Disabled

IP Address   192 . 168 . 1 . [20]

Figure 23-10   DMZ settings.

DMZ settings

There are certain circumstances where it is desirable to place a computer into the DMZ, where it will be directly exposed to the Internet. This router provides the option of placing a single computer into the DMZ, where it will also receive the public Internet WAN IP address. This is typically used for devices such as VoIP adapters, which must be exposed to the Internet to send and receive telephone calls. Placing a computer into the DMZ is not a decision to be casually made; exposure to the Internet can allow hackers direct access into devices.

Router review

Routers control the what, when, and "where to" of network communications, linking LAN devices such as computers and IP cameras to each other and the Internet. As such, understanding the control options of routers is an important component of IP-enabled security device installation. Routers also provide valuable firewall and network security features; it is the responsibility of the network administrator and/or electronic security company to properly enable them.

Gimme danger

Randomly changing router options in an effort to enable communications to IP security devices over the LAN or Internet will cause great frustration for alarm technicians and may also disable enterprise data communications functions for the client, which is a particularly distasteful turn of events.

Trust the Security Networking Institute on this issue; been there, blew it up. Luckily, it was our own network, so the damage and embarrassment were contained.

Here are some rules for working with a router, especially one that is also providing connectivity for the homeowner's or business' computer network.

1. After opening the router's programming screens, PRINT OUT each relevant page BEFORE MAKING ANY CHANGES . . . If things go horribly wrong, at least you can reset the client's router back to its original state, restoring communications.

2. Make any changes one at a time. After the change is complete, reset the router and try the communications that were not working previously. If it doesn't work, try another setting. Don't make 2 or more changes at the same time, as they may have a cascading effect on each other, and you likely won't remember what to undo in the heat of the installation.

3. Understand what you are changing. If you don't understand it, DON'T TOUCH IT. There are many esoteric options in routers that generally will not affect the communications of IP cameras and security devices. The fewer changes, the better.

4. Resetting the power on routers and IP network devices after programming changes will occasionally make things happen the way you'd like.

5. FACTORY DEFAULT WARNING—Resetting the factory defaults will most likely kill the client's computer network, garnering the alarm technician ugly looks and heated words. And the technician will have to reprogram the entire network's LAN IP address table, DHCP settings, etc. DO NOT RESET THE ROUTER'S DEFAULTS.

6. READ #1 AGAIN. This is absolutely critical.

Wi-Fi router review

Wi-Fi routers combine a Wi-Fi access point, network router, and often an unmanaged network switch into a single form factor. These low-cost devices provide a wide variety of settings and options, only some of which need to be manipulated to successfully connect an IP-addressed security camera, DVR, or other such components. Technicians should be cautious when making any changes to a client's router.

Ethernet Camera Guided Tour

The manufacturers of electronic security products are responding to the exciting opportunities of networking by producing many new cameras with Ethernet connection capability. In this section a typical network-enabled camera will be examined, concentrating on those aspects of networking that are common to most network cameras and may require manipulation by the security installer to achieve communications.

The Ethernet camera used for this example is a Panasonic MV-NM100, a compact and powerful device with many attractive features. The MV-NM100 has selections for all of the options usually available in network cameras, so it provides an excellent example for this tour.

Connecting to an Ethernet camera for initial programming

The two primary methods used to connect to an Ethernet camera are either to use the default IP address or to use the manufacturer's setup software, which will provide a "MAC" search function.

First a laptop or PC must be connected to the camera via a Cat 5 UTP jumper cable. If the camera is directly connected to the computer, a crossover cable must be used, whereas if the camera and computer are both connected to a gateway router, hub, or switch, the jumper used is a standard patch cable.

For initial programming, it is generally easier to use a directly connected computer rather than going through the network. For this example, the IP camera is directly connected using a crossover cable.

Connecting to the camera using default IP address

Every new IP camera is shipped with a preprogrammed default IP address from the factory. To reach the camera's programming options, first review the camera's instructions and locate the default IP address. The default IP address for the MV-NM100 on initial powerup is 192.168.0.10. After connecting the programming PC or laptop to the camera and connecting power to the MV-NM100, reset the "Local Area Connection" IP in the computer to an address on the same network as the camera. For this camera, the computer should be programmed to an unused address in this range, 192.168.0. (1-255). A simple address to use would be 192.168.0.1.

Open an Internet browser such as Internet Explorer, and type in http://192.168.0.10

The camera's username/password screen should appear. Use the proper entries obtained from the instruction manual. The control and programming fields of the camera are now available for review and changes.

Connecting to the camera using MAC search

Camera vendors often supply installation software that provides a MAC search function for connecting to cameras and setting the camera's IP address. In the case of the MV-NM100, after connecting the PC to the camera, inserting the installation disk will display the screen shown in Figure 24-1.

Figure 24-1   Panasonic MAC search entry screen.

When the "Setup" button is clicked, the screen shown in Figure 24-2 is the next one to appear. Manufacturers program pre-set MAC addresses into each IP camera or device. These are unique hardware identification codes; every individual device has a different MAC address. The installation software knows the range of MAC addresses used by that particular vendor, and will search out and display the related devices on the search screen.

In this case, a single camera has been located. Highlighting the MAC address and clicking "Network Setup" will open the screen shown in Figure 24-3, where the basic IP address information can be accessed and changed.

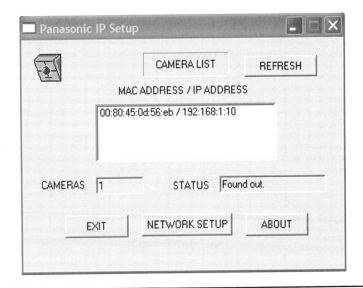

Figure 24-2 Panasonic search camera list.

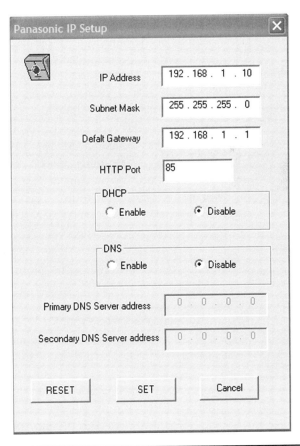

Figure 24-3 Panasonic search IP screen.

Notice in this example that the IP address of the camera is 192.168.1.10; this camera has already had its IP address changed to work with the Security Networking Institute's LAN.

Either of these methods can be used to access this particular camera's programming. Be aware that some Ethernet cameras only provide the default IP method of initial communication.

> **SECURITY TECHNICIAN'S NOTE:** Active software firewalls will often block MAC search programs. If the MAC addresses don't appear when searched, and the camera is powered and connected properly, turn off the software firewall and try again.

Programming network addressing

After using one of the two methods above, the camera is accessible for addressing and programming. The installer should be prepared at this point with the IP address that will be used by the camera, as well as a port selection if the camera is to be viewed from over the Internet.

IP addressing— MAC search

If using the MAC search function detailed above, it is a simple matter to change the IP address of the camera to its "permanent" setting. The camera illustrated in Diagram 3 above is programmed for IP address 192.168.1.10, standard subnet mask, Default Gateway (the gateway router's LAN IP) 192.168.1.1, and is set to Port 85. DHCP and DNS are disabled. The technician can make any changes, click "Set," and the changes will be made.

If the IP address has been changed from one Class C network to another, such as from the camera's default IP 192.168.0.10 to 192.168.1.10, the camera will disconnect from the programming PC once the change has been "Set." This disconnection occurred because the camera and the programming PC are now addressed for different networks. Now to reach the camera, it will have to be properly connected to the client's network and accessed using an Internet browser, inputting the camera's new/permanent IP address.

IP addressing— default IP connection

When connected to the camera using an Internet browser and the camera's default IP address, the IP addressing screen can be accessed by clicking to the "Advanced Settings" and "Network" screen.

In the case of the MV-NM100, the screen shown in Figure 24-4 is displayed.

Here is where the technician can input the proper LAN IP address, subnet mask, and default gateway, which is the address of the gateway router to which the camera is connected.

Under these settings is a selection for "Network Speed." As Ethernet has developed, there have been various subcategories of Ethernet, such as "Half Duplex" and "Full Duplex." Normally selecting "Auto" allows the IP camera to negotiate the best transmission protocol to use in its communications with the router or switch to which it is connected. This particular camera offers both 100 and 10 Mbps Ethernet,

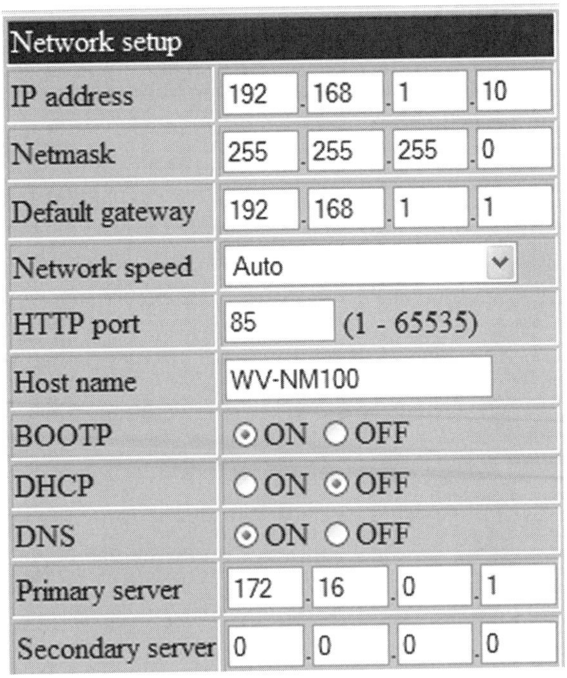

Network setup				
IP address	192	168	1	10
Netmask	255	255	255	0
Default gateway	192	168	1	1
Network speed	Auto			
HTTP port	85	(1 - 65535)		
Host name	WV-NM100			
BOOTP	⦿ ON ◯ OFF			
DHCP	◯ ON ⦿ OFF			
DNS	⦿ ON ◯ OFF			
Primary server	172	16	0	1
Secondary server	0	0	0	0

Figure 24-4 Panasonic IP setup screen.

while many of the currently available network cameras only operate at 10 Mbps. As mentioned previously, this can cause problems if switches, routers, or media converters that will be connected to the camera operate at 100 Mbps only.

Port address settings

The "HTTP Port" selection allows the technician to program camera access from the WAN or Internet onto a specific port, as was discussed in Chapter 9. This product, like many IP network cameras, shipped with the default port of "80," which is the port typically used by the network for HTTP ("Hyper Text Transfer Protocol") communications. HTTP connections require the host, in this case the camera, to connect to a viewing computer using a typical "web browser" program, such as Internet Explorer.

Host name

The "Host Name" can be selected to provide another name by which the device can be accessed over the network. For a camera, this name might be "North Lobby," "West Stairwell," or some other name that gives information about the view that the camera is providing to the monitoring computer.

BOOTP

Shorthand for "Bootstrap Protocol," BOOTP is a startup protocol that allows a diskless node, such as this camera, to determine its IP address and the address information for the gateway router to which it is connected.

DHCP

This setting allows the MV-NM100 to receive a Dynamic Host Control Protocol address from the gateway router. If this setting is turned on, the "Host Name" must be used to connect to the camera, as the IP address can periodically change.

DNS

Some larger networks will have their own internal DNS server, which resolves host names into their numeric IP addresses. If this particular camera is programmed for FTP image transfers and email notifications, and if the FTP and email servers are input as "Host names" in the "Alarms & Transmission" fields, DNS must be enabled and the DNS server address input in the field below.

Set up and get out

Once the proper settings have been input, clicking "Set" will apply any changes that have been made. If the settings are correct, the camera should now be accessible using a web browser from another network computer, provided that the camera has been properly connected to the gateway router.

Other network settings

While initially accessing the camera and setting the basic IP address information can provide a network camera with connectivity to the LAN, there are other features available that can be set to provide specific functionality to the device being programmed.

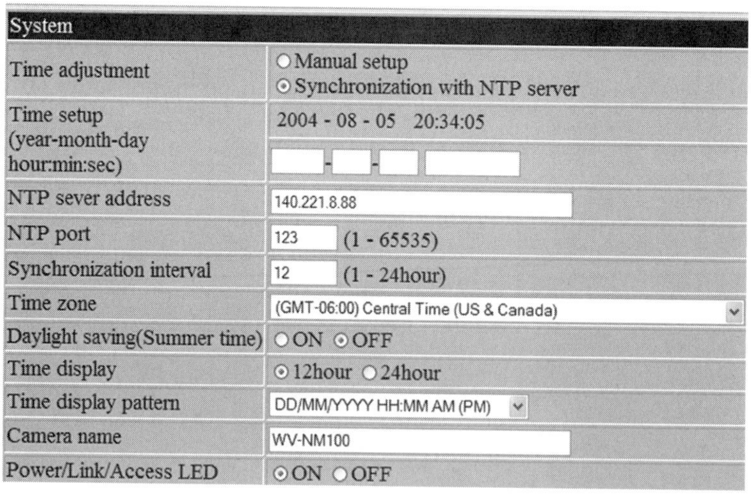

System	
Time adjustment	○ Manual setup ⊙ Synchronization with NTP server
Time setup (year-month-day hour:min:sec)	2004 - 08 - 05 20:34:05 　—　—
NTP sever address	140.221.8.88
NTP port	123　 (1 - 65535)
Synchronization interval	12　 (1 - 24hour)
Time zone	(GMT-06:00) Central Time (US & Canada)
Daylight saving(Summer time)	○ ON ⊙ OFF
Time display	⊙ 12hour ○ 24hour
Time display pattern	DD/MM/YYYY HH:MM AM (PM)
Camera name	WV-NM100
Power/Link/Access LED	⊙ ON ○ OFF

Figure 24-5 Panasonic system time settings.

NTP time settings

Time settings on IP cameras can be critical if the transmitted and stored images are to be used for evidentiary purposes.

If the camera is connected to the Internet, a Network Time Protocol (NTP) server's address can be programmed, which will automatically coordinate the time settings of the camera to the "atomic" clocks used to coordinate the time settings of Internet servers and devices. There are a number of different NTP websites available; generally this is a free service.

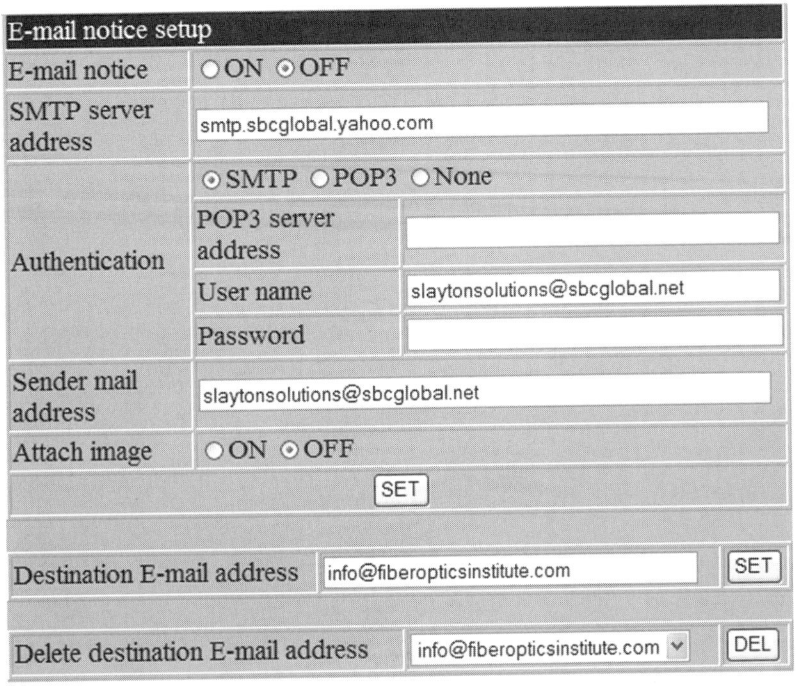

E-mail notice setup		
E-mail notice	○ ON ⊙ OFF	
SMTP server address	smtp.sbcglobal.yahoo.com	
Authentication	⊙ SMTP ○ POP3 ○ None	
	POP3 server address	
	User name	slaytonsolutions@sbcglobal.net
	Password	
Sender mail address	slaytonsolutions@sbcglobal.net	
Attach image	○ ON ⊙ OFF	
	SET	
Destination E-mail address	info@fiberopticsinstitute.com	SET
Delete destination E-mail address	info@fiberopticsinstitute.com ∨	DEL

Figure 24-6 Panasonic email settings.

The MV-NM100 can be programmed to transmit email notifications of alarm events, such as when motion is detected via the camera's internal motion detection program, or when an external device such as a door contact or alarm system motion detector that has been wired to the camera's external alarm input is activated.

In the programming fields, the settings for the type of email service, usually Simple Mail Transfer Protocol (SMTP) or Post Office Protocol version 3 (POP3) and "Authentication" can be obtained from the ISP.

Notice that there is a global "Email Notice" on/off setting at the top of the screen. This is a very useful option, as the email settings and motion detection areas can be set up in the camera, and email transmission can be quickly turned on or off by an authorized user. For example, the email transmission can be turned on at the close of business on Friday, and any motion detected over the weekend will generate email notifications. If the global email notice setting is turned off on Monday morning, then the recipient of the email notices will not be deluged with emails generated as people come and go during normal business times.

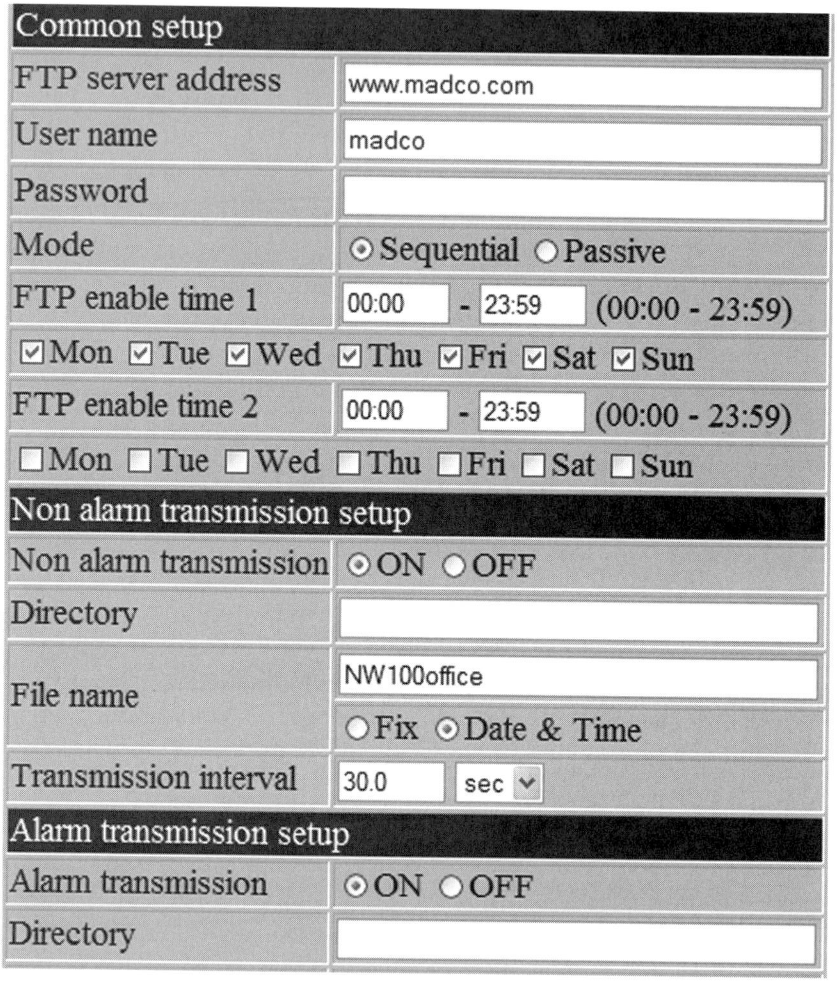

Common setup	
FTP server address	www.madco.com
User name	madco
Password	
Mode	⊙ Sequential ○ Passive
FTP enable time 1	00:00 - 23:59 (00:00 - 23:59)
	☑ Mon ☑ Tue ☑ Wed ☑ Thu ☑ Fri ☑ Sat ☑ Sun
FTP enable time 2	00:00 - 23:59 (00:00 - 23:59)
	☐ Mon ☐ Tue ☐ Wed ☐ Thu ☐ Fri ☐ Sat ☐ Sun
Non alarm transmission setup	
Non alarm transmission	⊙ ON ○ OFF
Directory	
File name	NW100office
	○ Fix ⊙ Date & Time
Transmission interval	30.0 sec ∨
Alarm transmission setup	
Alarm transmission	⊙ ON ○ OFF
Directory	

Figure 24-7 Panasonic FTP settings.

File transfer protocol settings

Shortened to FTP, this protocol is commonly used in Internet and network communications to send and access files. Many IP network cameras offer the option of FTP transfer, where an image can be transmitted to the FTP site either when motion or alarm inputs are activated or on a regular timed basis.

If the camera has accessibility to the Internet, the FTP website can be located remotely from the protected building. Also, FTP server software is available that can be installed into any PC on the LAN, providing image recording capability without investment in a new computer, Network Video Recorder (NVR), or DVR.

After inputting the FTP web address, username, and password, selections can be made as to when the FTP transmissions will occur and whether they are alarm generated and/or regularly timed. The filename and directory where transmitted files will be located on the FTP web site also can be input. When properly set up, individual image files will be transferred to the FTP site for review and retrieval.

> **SECURITY TECHNICIAN'S NOTE:** FTP setup can be tricky; it may take a few tries to get the proper settings in the camera. Also, there is a basic problem with using FTP to store images for review. Each image is an individual file, possessing a file name that is usually some combination of the camera's host name and the date/time of the image's recording. In order to review an incident, the operator will have to access the FTP site, input a functional username/password combination, locate the directory under which the images are stored, and start opening (and closing) the image files one at a time until the desired incident is located. This is a tedious and inexact process. If FTP is going to be used, it would probably be best suited for the storage of alarm images only, tied to the camera's motion detection or alarm inputs. That way, only the "alarm" images would need to be reviewed, not an entire day's images.

[Group A]		
Camera No.	IP address	Camera name
Camera1(Self)	192.168.1.10	WV-NM100
Camera2	192.168.1.11	WV-NM100
Camera3	67.95.109.49	SNIDemo
Camera4		

Figure 24-8 Panasonic multiscreen setup.

The MV-NM100 product offers a powerful multiscreen viewing function, allowing up to eight MV-NM100 cameras to be viewed on a single monitoring computer in two sets of four video feeds. Figure 24-8 shows the programming screen for the first set of four cameras; inputting the IP addresses and host names of the "other" cameras will place their images onto the "Quad Screen" display. Multi-view cameras can be connected to the LAN or from remote locations over the Internet.

Ethernet camera tour review

The primary concern of the alarm technician is the proper programming of an Ethernet camera's IP address and port number (if accessibility over the Internet is desired). Once the camera has been addressed and is reachable over the network, other settings such as video imaging issues, compression codecs, and image storage and transfer can be manipulated.

Remember, the features or options of various cameras and other networked security devices can differ widely as well as how common features are accessed or manipulated.

Wi-Fi Camera
Guided Tour

In the previous section we explored the setting options for a wired Ethernet network camera. The following section is a tour of a Wi-Fi wireless camera, and the differences and similarities between these two types of devices will be shown.

Why Wi-Fi?

Wireless cameras offer tremendous possibilities and options for electronic security installation companies to meet client needs. Because they're wireless, cameras can be quickly placed anywhere in the coverage area; just program the camera to the Wi-Fi network, plug in the transformer, and the camera is functional. The exponential growth of Wi-Fi networks in homes and businesses means that there are limitless opportunities to provide video services for existing and new clients.

Wi-Fi cameras are particularly adaptable for temporary covert surveillance. Wireless cameras can be programmed into the "ad hoc" mode (described later), so that the camera communicates directly with a Wi-Fi equipped laptop computer. The computer can now display the pictures and provide recording capabilities, both video and (with some specific Wi-Fi camera brands) audio. Just place the camera in the area to be monitored, aim the lens at the target area, and put the laptop in the room next door or down the hallway. By using an additional software package, motion detection can be activated, and recordings can be made when motion is detected.

Tour camera

The camera being used for this guided tour is a SOHO #CAS-230W/ANA available from Mobi Technologies. This inexpensive camera provides an excellent picture for indoor use, includes the lens, and provides both Wi-Fi and Ethernet communications.

> *Author's Note:* As with the previous tour of the MV-NM100 wired Ethernet camera, the SOHO Wi-Fi camera is supplied with a "MAC Search" programming software package, which provides a "Wizard" type of program for setting the basic camera communications parameters. To more fully illuminate the programming settings, this tour will be conducted by accessing the camera using Internet Explorer and the camera's default IP address.

Programming connections

Although Wi-Fi cameras are indeed wireless, initial programming must be done through some wired means. Wi-Fi communications requires that both ends of a network link must be programmed to the same SSID (Service Set Identification) and WEP (Wired Equivalent Privacy) encryption setting, if it is enabled. Camera vendors provide one of two methods to achieve this initial programming connection, either a serial port cable or an RJ-45 Ethernet connection.

In the case of the SOHO Wi-Fi camera, initial connection is via Ethernet, using either a crossover cable connected to a programming laptop or a standard UTP (see Chapter 6) Ethernet jumper if the camera is plugged into a gateway router, switch, or hub.

Once properly connected and powered, the camera is ready to be programmed. The technician must set the IP address of the programming laptop or PC to the same LAN network as the default IP address of the camera, which for all cameras of this make and model is 192.168.0.30. So the Ethernet IP address of the laptop must start with "192.168.0.XXX," with any number less than 256 and not "30" allowing communication with the camera.

Accessing the programming settings

Using Internet Explorer or another web browser program, type in the default IP address of the camera, and the now-familiar username/password screen appears. Consult the instructions for the specific camera for the default values to be filled in.

Wi-Fi settings

After entering the programming of the camera, a screen like the one shown in Figure 25-1 will be accessed.

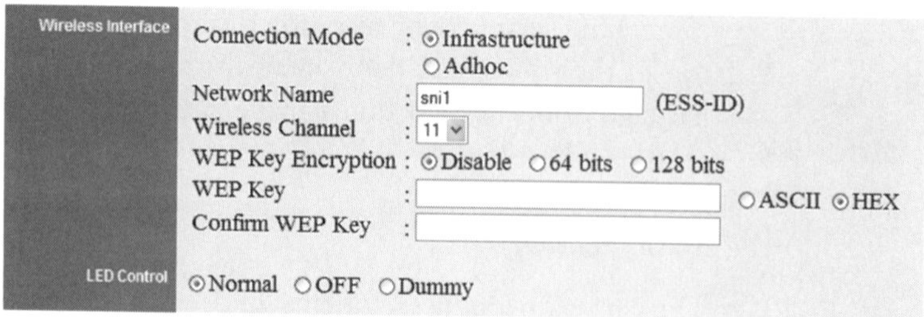

Figure 25-1 SOHO wireless settings.

Infrastructure or Ad Hoc

Here are the choices to be made to enable communication between the Wi-Fi camera and a Wi-Fi access point, router, or "ad hoc" programmed computer.

The first selection is "Connection Mode," where the technician can choose either "Infrastructure" (access point or router) or "Ad Hoc" (direct communication with another Wi-Fi equipped computer).

Network name

The "Network Name" is the SSID name that the particular network is using. This name must be the same as that programmed into the Wi-Fi router, access point, or ad hoc PC. This must be programmed exactly, matching any upper or lower case letters.

Wireless channel

There are eleven different channels in the 802.11b Wi-Fi spectrum. This channel setting should be the same as that of the access point/Wi-Fi router or ad hoc computer that will communicate with the camera.

Wi-Fi access points and routers will auto-negotiate the proper channel with devices and computers that have the correct SSID and WEP encryption (if enabled) programmed into them. If selecting the "ad hoc" mode, the user must manually program the channel being used in all devices that are to communicate together.

SECURITY TECHNICIAN'S NOTE: Most Wi-Fi products ship preprogrammed to Channel 6, and most users leave their devices on this default value. This situation means that the Channel 6 band becomes very crowded, particularly if there are a number of overlapping Wi-Fi networks in the same airspace, such as in a multi-unit retail or residential building. Use a different channel for the Wi-Fi camera, if possible. This will provide "cleaner" bandwidth with less interference for image transmissions.

WEP encryption

The next two fields provide for the selection of the WEP encryption type, and the specific WEP key being used by the network. As with the SSID name, this information may have to be extracted from the existing Wi-Fi access point or router so that it can be input into the camera.

LED Control

Below the WEP encryption fields is a selection for "LED Control," which allows the technician to disable the LEDs on the front of the camera, and is a wise idea for covert surveillance applications.

IP address settings

The screen shown in Figure 25-2 provides the input information for the IP address of the camera.

Camera Name	CS-01CE91
Location	
Admin	Admin ID : sni1
	Admin Password : ••••••••
	Confirm Password : ••••••••
IP Assignment	⊙ Manually Assign
	IP Address : 192.168.2.21
	Subnet Mask : 255.255.255.0
	Default Gateway : 192.168.2.1

Figure 25-2 SOHO IP settings.

At this location the username and password can be changed and the IP address set. Remember that each device on a LAN must have a unique IP address, whether Ethernet or Wi-Fi. The Wi-Fi device cannot have the exact same IP address as an Ethernet device.

Selections for the "subnet mask" and "default gateway" are also made in this programming area.

Port settings

The screen image in Figure 25-3 shows where the port settings for the SOHO camera can be changed.

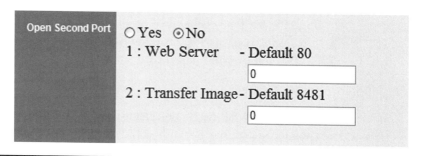

Open Second Port	○ Yes ⊙ No
	1 : Web Server - Default 80
	0
	2 : Transfer Image- Default 8481
	0

Figure 25-3 SOHO port settings.

As with most Ethernet or Wi-Fi IP cameras, the default port is "80," which is the typical port used for HTTP communications. Notice that this camera, like some others, actually uses two ports; one for web browser communication, and one for transmission of the images. In the event that this camera is to be viewed from over the Internet, both ports (or both port numbers selected) must be cleared through the gateway router and ISP adapter (cable or DSL, if used) so that communications can travel from the Internet to the camera and vice versa.

DNS and DDNS settings

Like other IP network cameras, the SOHO provides the ability to program a Domain Name Server (DNS) IP address into the camera. This may be necessary if the camera's automatic email or FTP image storage is activated.

Dynamic Domain Name Service (DDNS) is a combination of client software and an Internet server that will track the changes of a DHCP dynamic IP address, as is typical with lower cost cable and DSL ISP services. The SOHO Wi-Fi camera provides software that will perform the "uplink" transmission of the IP address to the DDNS service, provided that the fields shown are properly programmed.

Set and get out

As with most IP-addressed devices, once changes have been input into any of the programming fields, the changes must be "set" or "saved" to cause them to be input into the device. After making the proper programming selections, the technician will click the "save" button to implement the programming changes.

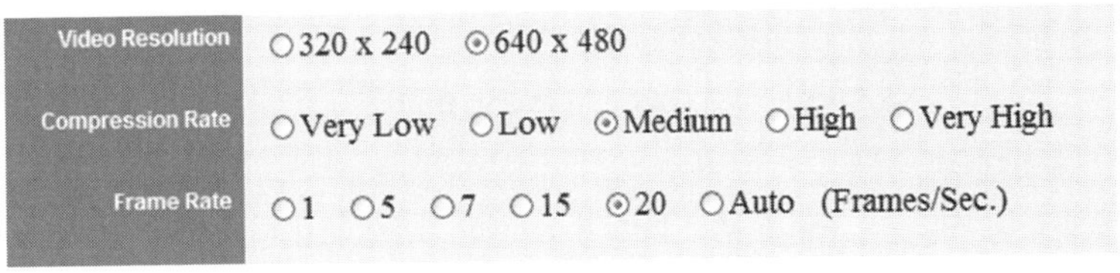

Figure 25-4 SOHO image settings.

Image settings

The SOHO Wi-Fi camera provides a simple-to-use "image settings" selection screen. Two options for "Video Resolution" or scaling are available, either 320 × 240 or 640 × 480. The "compression rate" can also be selected, with "very high" compression resulting in smaller file sizes, faster frame rates, and lower quality of image. The "frame rate" is also selectable, with the "auto" setting providing the highest frames per second that the particular network connection is capable of handling at any given time.

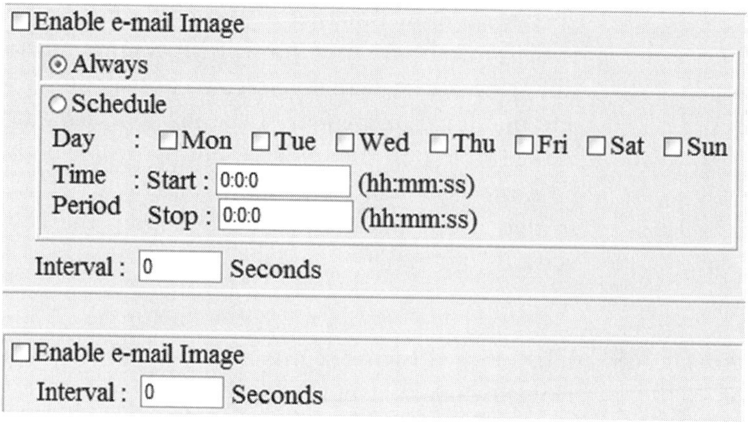

Figure 25-5 SOHO email settings.

Email settings

Different products provide various capabilities for the transmission of images from a network camera. In the case of the SOHO Wi-Fi camera, it can be programmed to email a string of images based on the day of the week and a programmed time period. This feature can be used to transmit images during off-hours, weekends, or whatever time and date schedule is selected.

SECURITY TECHNICIAN'S NOTE: Using a camera's email functions can provide a backup image "recording" function, even if the camera's video output is being simultaneously recorded by some other means. Also, email imaging provides the additional security of transmitting the camera's images to an off-site location, where they are inherently protected against the intruder who might attempt to disconnect or destroy a local Network Video Recorder (NVR).

Wi-Fi camera review

In order to program a Wi-Fi network camera, it must be initially accessed through some wired means, either through a Cat 5 UTP connection or a serial port hookup.

To enable wireless transmission, the technician must know the SSID, channel number, and WEP encryption code for the access point or Wi-Fi router that will provide the communications link. This information must be programmed properly into the Wi-Fi camera.

As with wired Ethernet cameras, different products will provide different features for the viewing and transmission of images.

CHAPTER 26

Wireless Laptop Surveillance

By programming a Wi-Fi-enabled network camera in an ad hoc network with a Wi-Fi-equipped laptop, a covert video recording setup can be installed and activated in a matter of minutes.

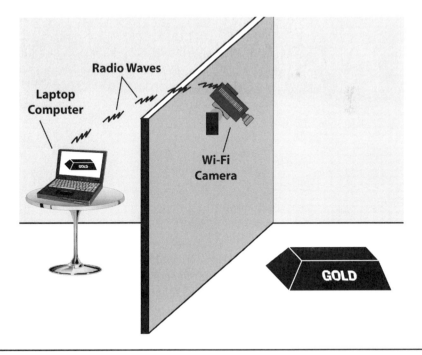

Figure 26-1 Wi-Fi surveillance.

Equipment list

1. Wi-Fi network camera(s), power supply (pack and cord), programming/connection cable (for this example, the camera is a Veo Wi-Fi "Observer").

   Note: The camera must be able to function in ad hoc Wi-Fi mode. Not all products provide this capability.

2. Laptop computer, Windows-equipped, with Wi-Fi built-in or separate Wi-Fi card.

3. Video recording software—may be supplied with the camera, or a third-party software package such as "ReCam" can be used. This software must be pre-installed into the laptop being used as a DVR.

SUGGESTION: The author strongly suggests that the technician perform the following setup and testing prior to taking the equipment to the client's location. By setting up the system beforehand and confirming functionality, the actual installation can be performed rapidly and without difficulty.

Camera programming

The camera must be programmed for the following:

1. IP Address—To enable network communications with the laptop
2. Wi-Fi Ad Hoc Mode—To enable wireless communications between the camera and laptop

Accessing the camera's programming fields

In general, Wi-Fi-enabled network cameras require a wired connection for initial programming. This connection may be accomplished with a Cat 5 RJ-45 "crossover" cable between the laptop and the camera, or there may be a programming cable included with the camera that provides for a serial or USB connection. Consult the camera's directions for the proper programming connection method.

WARNING #1: Some laptops no longer have a 9-pin "D" serial connection built in. If such a serial connection is required, and if the laptop doesn't have a standard serial port, a USB-to-serial converter can be used to achieve the needed programming connection. Some manipulation of the Hardware Device Manager (Windows XP) may be necessary to activate the USB to serial connection. Consult the instructions included with the converter.

WARNING #2: A laptop's "firewall" software may block communications to the camera via a serial or Cat 5 connection cable. Unlike blocked communications coming over the Internet, the laptop's firewall may block communications to the camera without displaying a "warning" or "communications blocked" pop-up. Disable any firewall software before attempting to access the camera's programming fields.

SECURITY TECHNICIAN'S NOTE: If using a wired Ethernet camera for connection to a laptop, use a "crossover" cable to allow communications between the two devices without having to employ a hub or switch.

Setting the camera's IP address

Once the camera's programming is accessed, the IP address of the camera must be set so that it and the corresponding laptop are addressed for the same Class C type of network (See Chapter 9 for more information on Class C networks). A simple way to perform this is to check the Wi-Fi IP address of the laptop (see Chapter 11). This address will typically be in the format of

192.168.X.YYY

with X being 1 or 0, and YYY being a value from 0 to 255. Once this information is retrieved, program the IP address of the camera to a corresponding Class C address. Some examples of functional address sets:

Laptop 192.168.1.12/Camera 192.168.1.5
Laptop 192.168.0.12/Camera 192.168.0.5

Programming the camera for ad hoc Wi-Fi mode

An ad hoc Wi-Fi setup allows programmed devices to communicate with each other without the need for a separate Wi-Fi router or access point.

Programming of the ad hoc mode is performed by three selections in the camera's programming fields. The first is to select the "ad hoc" connection type. The second is to input an SSID "name" into the appropriate field, and the third is to select a Wi-Fi channel for the ad hoc devices to communicate with each other. Figure 26-2 shows how it's done on a Veo Observer.

This screen also provides fields for the IP address settings.

You may or may not choose to set the WEP encryption for this particular application. If the camera and laptop recorder are to be set up temporarily, not activating the WEP will provide higher frames per second and less trouble during programming and installation.

Once the ad hoc settings are input, press "Save Settings" to complete the programming of the camera.

> **IMPORTANT:** Leave the camera powered, and physically close to the laptop. The camera must be powered and programmed to complete the programming of the laptop.

Programming the laptop for ad hoc mode

Three settings in the laptop must match those of the camera for the devices to communicate (four if you are using WEP encryption). They are:

1. The IP address and subnet mask must match the network Class of the camera

2. Ad hoc mode must be selected

3. The SSID of the ad hoc network must match

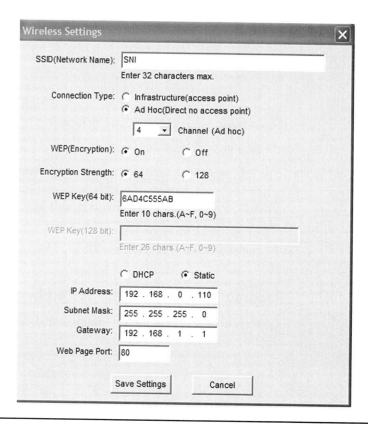

Figure 26-2 Wireless laptop Veo settings.

These programming fields are standard among the vendors of Wi-Fi communication cards, although there may be differences in how they are accessed.

If the laptop has built-in Wi-Fi, the program fields are accessed through Windows. To find and change these settings in Windows XP, from the desktop:

1. Click "Start."

2. Click "Control Panel."

3. Click "Network and Internet Connections."

4. Click "Network Connections."

5. Right-click "Wireless Network Connection."

   (If there is no "Wireless Network Connection" icon within the "Network Connections" window, the software drivers and controls for the Wi-Fi card are located in another software section of the computer.)

6. Highlight and click "Properties." You will see a window like the one shown in Figure 26-3.

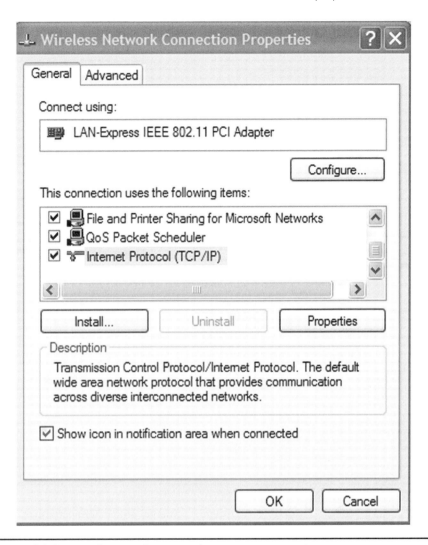

Figure 26-3 Windows wireless network connection properties.

Scroll down on the right side of the center until you see "Internet Protocol (TCP/IP)". Highlight this and click the "Properties" button. You have reached the "Internet Protocol (TCP/IP) Properties" window, as shown in Figure 26-4.

Here you select "use the following IP address," and input an IP address that is in the same class as what you've programmed into the camera, i.e.:

Camera IP: 192.168.1.110
Laptop IP: 192.168.1.12

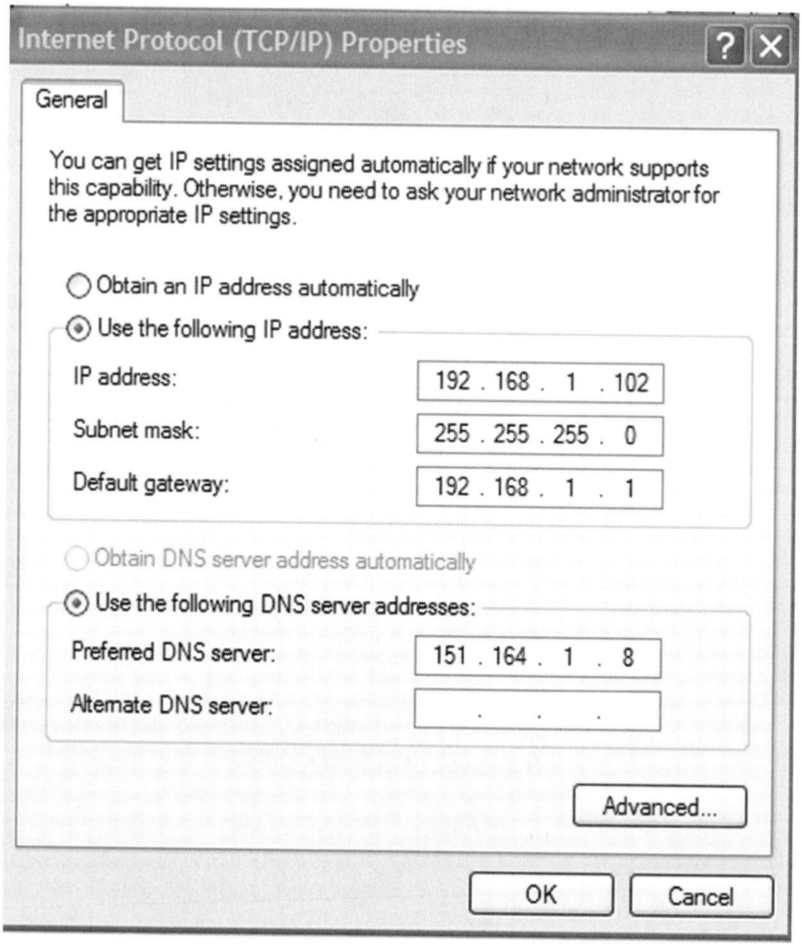

Figure 26-4 Windows internet protocol properties.

As there is no default gateway, and the network will not be accessing the Internet, settings for the ""preferred DNS," "alternate DNS," and "default gateway" are not required. Click "OK" upon completion of the IP settings.

7. Click the "Wireless Networks" tab on the "Wireless Connection Properties" window. A screen like the one shown in Figure 26-5 will pop up.

8. Click the "Advanced" button in the lower right. This will take you to your choices for the type of Wi-Fi network that is desired.

9. Click "Computer to Computer (ad hoc) networks only," and "Close."

10. Back at the "Wireless Network Connection Properties" window, you should see the SSID name that you programmed into the camera in the upper "Available Networks" window. Highlight the SSID and click "Configure."

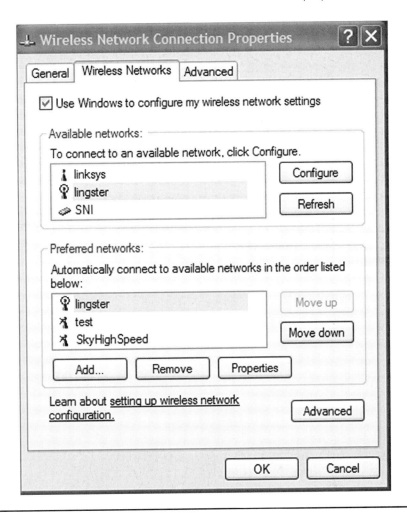

Figure 26-5 Windows wireless network tab.

In the "Association" window, the previously selected SSID will be displayed, and the only option available to the installer will be to indicate whether or not WEP has been selected in the camera. If so, click the "Data Encryption" box and input the eight characters (1–15, A–F) that constitute the WEP code.

> **WARNING:** As was reviewed in the previous Wi-Fi section, not all WEP encoding is the same . . . some 64-bit devices accept a 10-character code, others accept eight. It is very possible, based on the camera selected and the Wi-Fi card/software in the laptop, that WEP cannot be enabled for a particular camera and laptop setup.

The author recommends setting up your surveillance camera and laptop without WEP enabled. After establishing that everything is working properly, the installer can revisit the programming of the camera and laptop and then attempt to set up WEP encryption.

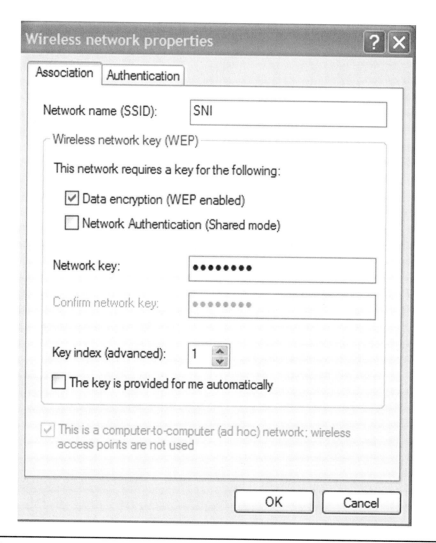

Figure 26-6 Windows wireless connection properties.

11. After completing the WEP selection, Click "OK" and close out of the "Wireless Network Properties" windows until you return to the main "Network Connections" window. Now you want to "Enable" the "Wireless Network Connection" icon by right-clicking on its name, highlighting "Enable," and clicking.

12. Close out of the "Network Connections" window.

13. From the Desktop, your Network icon should be displayed, probably in the lowest right corner. Float your cursor arrow over the icon and a balloon will appear, showing the connection, data speed, and signal strength.

14. To access the camera, open your web browser and type the camera's IP address into the address line. (Turning off any other network connections will simplify the laptop's search for the ad hoc camera.)

Reach the sign-on screen and input the camera's default username and password. You are now accessing the camera's images and functions using a Wi-Fi ad hoc network.

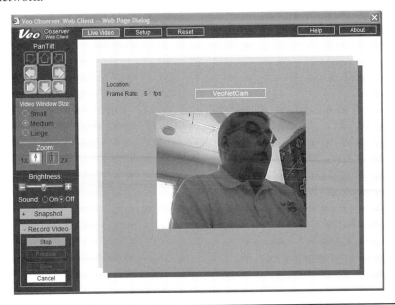

Figure 26-7 Wireless laptop Veo live video screen.

Security issues

Remember to change the username and password of the camera and to deactivate the camera's "transmission" light.

"Hey, it's not working!"

Try and/or check the following:

1. The laptop's "Wireless Network" is enabled.
2. The SSID names match between the camera and the laptop.
3. If WEP is activated, the encryption codes match between the devices (if there's a doubt, turn off the WEP in both devices).
4. The laptop and camera are both set for the "Ad Hoc" mode.
5. Try toggling the Wi-Fi network off and on. "Disable" then "Enable" the "Wireless Network Connection."

Recording software

Network cameras function as web servers, allowing viewing of the camera's images by authorized users. To record those images requires a separate software set to be installed in the laptop. This software may have been provided with the camera, which is the case with Veo and some other brands. Third-party software is also available that can perform this recording function.

Disk space requirement

The laptop must have sufficient unused disk space to record for the amount of time that is required. To check for unused disk space:

1. Click "Start."
2. Click "My Computer."
3. Under "Hard Drive Disks," right-click the "Local Disk" (usually drive C).
4. Highlight "Properties."

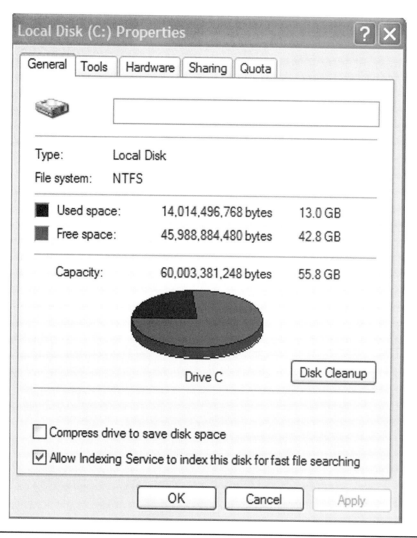

Figure 26-8 Windows local disk properties.

A window like the one in Figure 26-8 will indicate the amount of unused space available on the hard drive.

Testing for video storage needs

A simple way to test for disk space requirements is to set up and program the camera and laptop at your office before installation at the client's location. Activate recording, and let the system record video for ten minutes.

Stop the recording, and access the recorded file. Open "Properties" in the file header and locate the size of the file, which will usually be listed in megabits (MB). For example, Figure 26-9 shows the "Properties" screen for an 11-minute recording from a Veo Observer Wi-Fi camera.

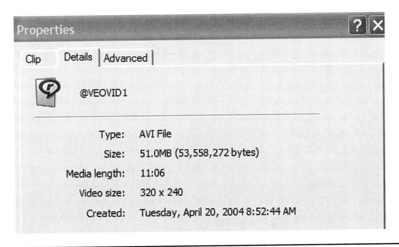

Figure 26-9   Wireless laptop Veo video file properties.

Multiplying this value by 6 will provide an accurate approximation of the amount of disk space needed to record one hour of surveillance. In this case, one hour of recording will equal approximately 300 MB for one hour, or 7.2 GB for 24 hours. When calculating the disk space requirement, don't forget about weekends or other times when the laptop won't be routinely accessed and the file viewed or saved onto CD-ROMs.

Onsite installation

After pre-programming and testing the Wi-Fi camera and laptop, installation of the system for the client should be quick and easy. Here are the steps:

1. Pre-charge the battery in the laptop—it will be used to verify signal strength, and the technician may have to move the computer to a location where signal strength is adequate.

2. Place the camera in the location to be viewed and apply power.

3. Turn on the laptop and establish communications with the camera (view the image).

4. Move or adjust the camera to achieve the desired view.

5. Leaving the camera area, carry the laptop to its proposed location. This area should be as close as possible to the camera, while providing an AC power outlet and physical security. An adjoining office or locked storeroom would be optimal.

6. Confirm that the camera's signal reaches the laptop. Check the signal strength by moving the laptop's cursor arrow over the "Network" icon in the lower part of the desktop. If using the Window's XP Wi-Fi software, the signal strength should be "Good" or "Very Good."

7. Plug the laptop's power pack into an unswitched AC outlet and into the laptop itself.

8. Confirm that the computer has adequate unused hard drive space to record for the time frame planned.

9. Instruct the client on how to retrieve, view, and store the recorded video. Some recording software is protected by username and password entry, which will need to be set.

10. Leave the computer turned on. Power-saving features of the laptop may need to be turned off. The laptop may go into "Standby" mode, but recording should not be hampered.

Wi-Fi warning

Wi-Fi wireless communications operate on "line of sight," and can "drop out" for a number of reasons. If, for example, the camera is on one side of a hallway in an office, the laptop/recorder is on the other, and a large metallic object such as a rolling box or shelf is placed so as to block the signal, the recording may be stopped. If such a stoppage occurs, there will be no immediate notification to the client. Careful placement of the camera and consideration of the radio signal path can reduce or eliminate this potential problem.

Wi-Fi signals can also be disturbed by interference from other devices. Verify that the signal strength is "good" or "very good" between the laptop and the camera. Allow some time to test the system, perhaps one half hour, while the technician is onsite before turning it over to the client.

Suggestions

1. Program and test the system before taking it to the jobsite.

2. At the jobsite, place the camera and laptop as close as possible to each other without compromising security.

3. Disable or do not turn on the WEP for short-term installations.

4. Set the camera for the lowest resolution that provides an adequate image.

5. Check for adequate Wi-Fi signal strength.

6. Verify that the laptop has sufficient unused hard drive space to record the amount of video needed.

7. Carefully instruct clients in the uses of the system, particularly for the transfer to CD of any video segments they are interested in retaining.

8. Have the technician's office or central station place scheduled telephone calls to the client reminding him/her to view the video that has been previously recorded.

Summary

A laptop PC and a Wi-Fi enabled security camera can be programmed to communicate directly with each other, using the "ad hoc" method. Once the two devices are programmed, the laptop can provide a DVR function. There are two critical issues: first, the signal strength between the two devices should be tested to verify that a sufficient communication path has been established. The second issue is that there is adequate space on the laptop's hard drive to store the desired elapsed time of video transmission.

CHAPTER 27

IP Video Server Guided Tour

Video servers provide wired Ethernet network connections for existing (or new) analog output CCTV cameras. Using standard BNC connections, any CCTV camera can be connected to an Ethernet LAN and/or the Internet. Video servers will accept the analog video input, apply various types and levels of data compression, and transmit the video signals to authorized users or preprogrammed software packages installed on a local or remote PC. In this section, a typical IP network video server will be viewed, while highlighting programming options for installing technicians.

The product used for this guided tour is a BlueNetVideo model BB01S. This particular device has great potential for alarm company use, as it is relatively low-cost, uses standard BNC connections for the camera, and provides a "pass through" BNC output along with the camera input. This last feature is very important, as it enables the BlueNetVideo server to retain the video signal feed to the local monitor when it is connected to an existing CCTV system. This particular product also ships with the BlueNetVideo "I-Pro" software, which allows simultaneous viewing of up to sixteen video feeds from BlueNetVideo servers and provides NVR (Network Video Recorder) functions.

Programming access

As with other devices included in this "guided tour" section, technicians can use a desktop or laptop PC to view or change programmed options. The computer must be programmed to the correct Class C IP network address, and connected to the video server with a UTP Cat 5 jumper. If connected directly to a laptop or PC, a crossover cable must be used; standard cables are used if the video server is connected through a gateway router, switch, or hub.

IP addressing options

At this point, the selections for IP and port addressing should be familiar to the reader. Figure 27-1 shows the IP addressing screen for the video server.

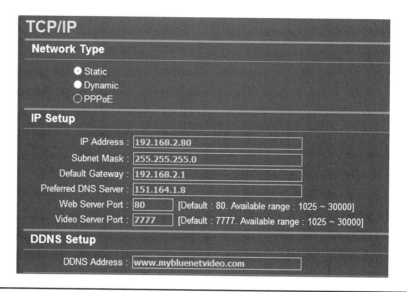

Figure 27-1   BlueNet IP port settings.

The typical settings options are available, such as static or dynamic IP, IP address, subnet mask, default gateway, etc. As with the Wi-Fi camera used for this volume's "guided tour," this video server uses two ports for communication, one for transmission of video images, and the other to communicate with the video server's web functions.

DDNS functions

BlueNetVideo provides its own connection to a DDNS server, which the installer may select to use, or a different DDNS service may be programmed into the video server.

Image settings

The screen illustration Figure 27-2 shows how the selections for the video image can be manipulated.

This particular server provides on-screen captions, which can be quite helpful when viewing multiple cameras simultaneously. Selections can be made for the camera name, date/time, frames per second, and other caption options.

Below the camera name and caption selections are the settings for scaling, black and white ("gray") or color ("rgb") camera input, as well as "CCD Assembly" and "Camera Mount" selections. These last two selections will invert the transmitted image so that a ceiling-mounted camera provides a "right side up" image even if the camera is technically upside down.

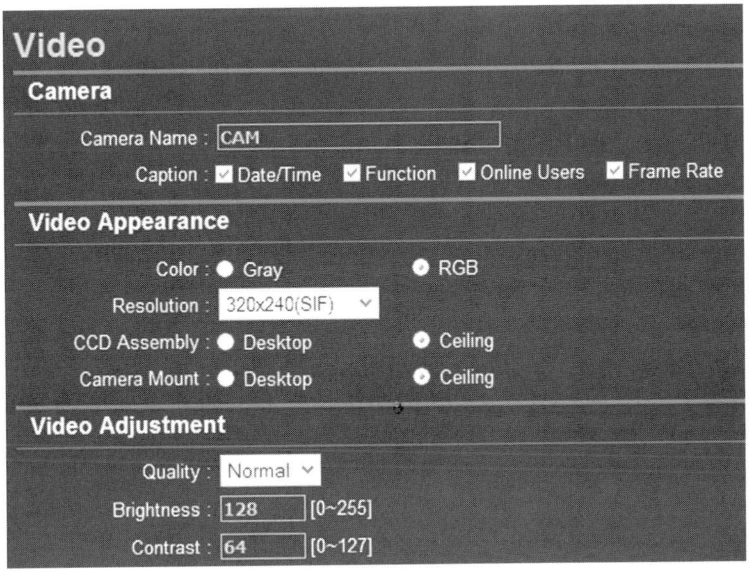

Figure 27-2 BlueNet image settings.

Notice the settings for "Video Adjustment." The pull-down window next to the "Quality" heading provides a variety of selections, from "Fastest" to "Best." This setting is actually the compression percentage that is being applied to the individual images. "Fastest" will be the most compressed, while "Best" provides the highest quality picture, at a reduced fps rate in comparison with the "Fastest" setting.

SECURITY TECHNICIAN'S NOTE: It is important to note that a video server will generally employ one of the popular compression formats. The BlueNetVideo BB01S, for example, uses MJPEG (motion JPEG) as its compression algorithm. The pros and cons of MJPEG and MPEG are detailed in Chapter 17. Issues are raised when more than one video server is going to be installed, for example in a number of retail stores that are to have their video images viewed at one central location. What is important to consider is what software will be used to view and record the multiple images; the software selected must be compatible with the compression format(s) coming from the remote video servers. This applies to both the current installation and any new ones that may be added to the network. In general, the electronic security contractor must pick a video server software combination and stick with it, at least for that particular client. Changing server vendors may not be possible, as the "new" video servers may use a different compression codec that cannot be processed by the existing monitoring program.

Video image options

As a fairly simple IP video server, the BlueNet device offers limited image settings—only "brightness" and "contrast."

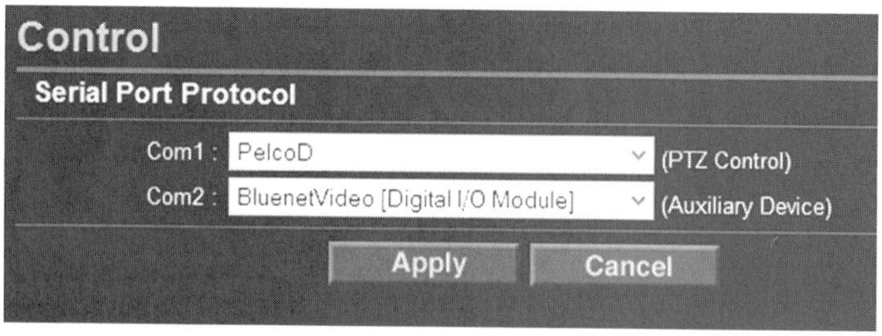

Figure 27-3   BlueNet serial port settings.

Serial port settings

If a camera has pan-tilt-zoom ("p/t/z") functions, those capabilities may need to be controllable through the IP video server. The BlueNetVideo BB01S provides an RS-232 output port that can be connected to the control circuitry of the p/t/z mechanism to enable control by authorized users. The same output port on the video server can also be programmed for Auxiliary In/Out functions.

SECURITY TECHNICIAN'S NOTE: To enable p/t/z functions through a network video server, the server selected must support the control signal protocol that is used by the particular p/t/z device being connected. Most video server manufacturers will support most major-brand protocols; however, it would be wise to confirm such support before purchasing and attempting to connect a p/t/z camera to a video server.

Video server tour summary

Video servers provide wired Ethernet and Internet connections for traditional analog CCTV cameras, compressing the video images and transmitting them to authorized users. While IP and port addressing and other selections are the same for IP and Wi-Fi addressed cameras, video servers generally have more options for picture quality, frame rate, and other image issues.

Video servers are perfect for existing CCTV systems that need remote network viewing capabilities; a single video server can take the "quad" image from a monitor and allow remote viewing of all four cameras simultaneously. Using video servers also allows the electronic security company to use any analog CCTV camera, including outdoor and/or p/t/z cameras. Such functionality may not be available in a camera that has built-in IP capabilities.

Video Management and Recording Software Guided Tour

Video streams from IP network cameras and video servers can be viewed, controlled, and directed to digital storage locations through the use of a video management software package. One such powerful and inexpensive product is called "ReCam," and will be used as the example for this guided tour. In this section the common programming selections for technicians will be detailed along with some of the powerful features available in video management software.

Why use video management software?

Basically, a video management software package automates many of the functions of camera control, monitoring, and recording that, without management software, have to be manually controlled for each camera being monitored.

Video management software provides an automatic platform for video issues, reducing or eliminating operator input. Once the video management software is configured, recording of video (and in some cases, accompanying audio) images is automatic, along with other features such as FTP storage, email notifications, and motion detection. This automation is highly preferable to depending on individual users to "turn on" the recording of a network camera.

Software details

ReCam software from Mobi Technologies operates on Windows-equipped PCs, and will process most IP cameras and video servers that use MJPEG or JPEG compression algorithms. Along with a host of valuable camera features, ReCam also provides Network Video Recorder (NVR) functionality, directing the storage of video images onto the monitoring computer or anywhere else that storage space is available on the LAN.

Program startup

Once the software has been installed into the monitoring PC, the user simply clicks the ReCam icon on the desktop to start the program.

Camera connection

The camera programming "wizard" included with ReCam will prompt the technician through the steps necessary to add a new camera or reconfigure an existing IP-enabled video device.

In Figure 28-1, after selecting the "Network IP Camera" radio button, the operator is prompted to select the manufacturer/part number of the camera being connected. As different cameras can employ slightly different settings, it's helpful to have these pre-set in the software.

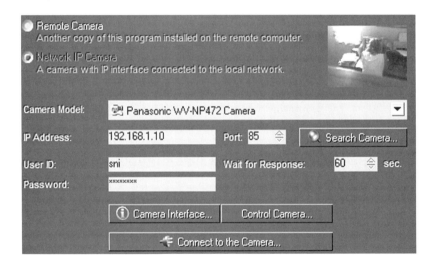

Figure 28-1   ReCam IP camera setup.

After selecting the camera type, the IP address and port number are input. Notice that the software asks for the username and password of that particular camera; once programmed, the ReCam software will automatically connect to the programmed camera and input the correct user information. This is a powerful security feature, as the passwords allowing access to an individual camera's programming features can be retained by the system administrator and not given out to the monitoring personnel or guards.

Clicking "Connect to the Camera" will test the settings that have been input; if correct, the ReCam software will connect to the selected camera in a minute or less.

Video and audio compression

Once the camera has been connected to the software, a variety of options are available to direct and control the video and audio streams' recording functions.

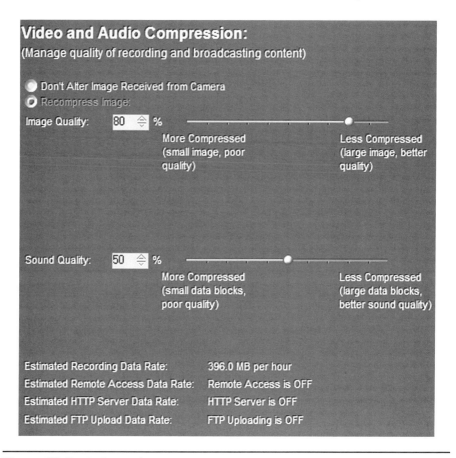

Figure 28-2   ReCam video compression screen.

Although the MJPEG or JPEG images that are received by the ReCam-equipped monitoring computer are already compressed, this software provides for further compression of recorded images, primarily to save storage space. In this example, the stored images will be compressed to 80% of their original received size.

Notice that at the bottom of the screen there is a display indicating the "Estimated Recording Data Rate." This is very helpful, as it enables storage requirements for video images to be accurately estimated.

It's important to note that the recording settings for the ReCam software are individually selected for each camera. Some cameras may be recorded at a higher rate than others, or some cameras may be recorded on a 24/7 basis while others may only be recorded when motion has been detected.

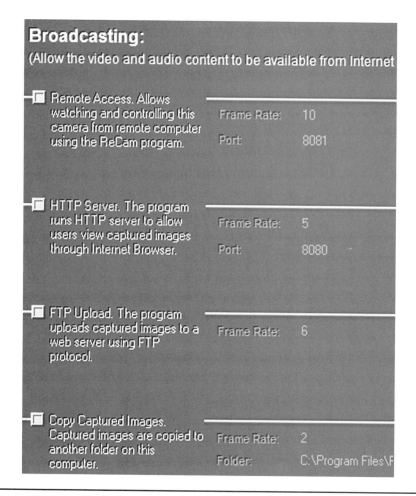

Figure 28-3 ReCam broadcasting settings.

ReCam provides a variety of different ways that video images can be viewed by the primary monitoring computer and other computers. In the "Broadcasting" screen, selections can be made for the following options:

- "Remote Access" allows other computers equipped with ReCam software to view selected camera images over the network or Internet.

- The "HTTP Server" option can provide video image viewing to authorized users connecting to the primary monitoring computer using common web browser software. This can be useful for "public" cameras; for example, at a child care facility that will be watched by parents.

ReCam also provides an FTP uploading service, transferring recorded images to an FTP server, either on the LAN or somewhere on the Internet.

The "Copy Captured Images" setting works when images are manually captured, storing such images into a specific directory and folder on the network.

Image scaling and cropping

Options are provided to "scale" captured images, which can further reduce file sizes for storage. Another option is "cropping," where the internal box on the screen can be adjusted and moved. What's inside the box will be captured, what's outside will be discarded. This is quite useful in a retail video situation, where the image includes the cash register as well as the overall front area of a store. While the entire image can be viewed, only the cash register portion is recorded.

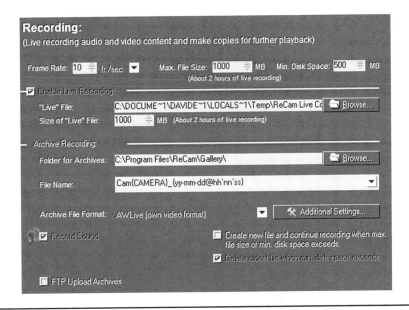

Figure 28-4 ReCam recording screen.

Recording options

This section of the programming details the network recording options available with ReCam.

The top line of selections includes "Frame Rate," "Max. File Size," and "Min. Disk Space." These settings are related to the "archive," or scheduled recording functions, which can be set for either specific dates and times or activated by motion detection. While the video stream from a camera may be upward of 20 fps, the operator can set the recording rate for fewer frames per second if desired. The "Max. File Size" defines how large an individual file of stored images can become before ReCam will start a new file. The "Min. Disk Space" setting ensures that the ReCam recording will not completely fill the available space on the drive that is storing the video images.

When the "Enable Live Recording" box is checked, "live" recording is active as long as the software and/or camera has not been turned off. This live recording can be viewed from the monitoring screen so that recorded video can be reviewed while the software is also recording real-time live images. The software provides a "browse" button, to select where on the network the recorded video files will be stored. It's important to note that the recording doesn't have to be stored within the monitoring computer; it can be placed anywhere that is accessible on the network.

Recorded file encryption

This software program can be set to apply encryption to the stored video images, allowing playback viewing only to those using the correct password.

Figure 28-5   ReCam encryption screen.

In Figure 28-5, 448 bit "Blowfish" encryption has been selected to protect the stored video images from viewing by unauthorized personnel or computer hackers.

SECURITY TECHNICIAN'S NOTE: Setting encryption for video storage will increase the size of the stored file.

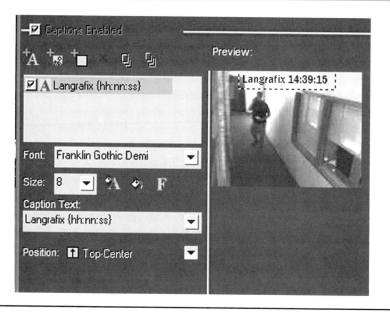

Figure 28-6   Captions.

Caption selections

Captions, including the time, date, business name, camera location, and other options, can be programmed individually for each camera. ReCam provides adjustable font styles and sizes, colors, opacity, and other options, and captions can be located anywhere on the image screen. Adding captions to an image will increase the size of the captured files and storage requirements.

Motion detection

While some IP network cameras provide on-board motion detection, more flexibility is available when using a video management software program such as ReCam.

Figure 28-7 shows the primary "Motion Detection" setting screen.

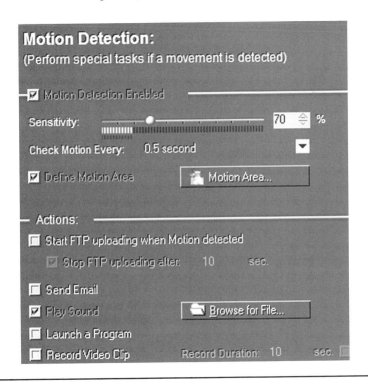

Figure 28-7 ReCam motion settings.

These are the selections as to whether motion is activated, the percentage of motion (sensitivity), and the actions that the software is to perform when motion is detected. Upon the detection of motion in the image, video clips can be recorded, the monitoring computer can play a sound file, FTP image storage can be activated, and emails can be transmitted.

The field of motion is set using an adjustable/moveable "box" overlay. This software product also provides the ability to upgrade the recorded fps and stored image quality after motion has been detected.

218 Chapter 28

Scheduling

Consider a typical commercial scenario, with multiple cameras providing views of different facets of a business. While one camera is watching the public entrance and waiting room, another is focused on the loading dock, another on the assembly area, etc. When setting up the video management software parameters, it would be attractive to have a simple screen that allows authorized users to schedule by camera when recording, motion detection, remote viewing, and other programmable features are operative. ReCam provides an easy to use "Scheduler" screen that allows each camera to be scheduled by the time and day of the week for particular recording and/or notification features.

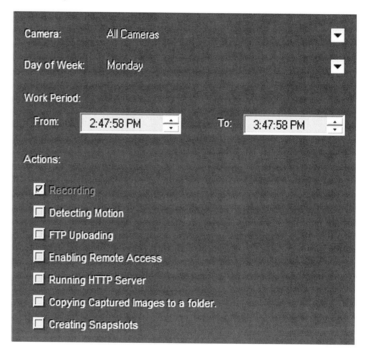

Figure 28-8 ReCam scheduler.

This is a very powerful tool to reduce the size requirements for image storage. For example, the camera viewing the public entrance may be set to record just a few frames per second during the typical workday, but to have motion-activated and high-frame recording during off-hours. When properly planned and programmed, the video information that is critical can be recorded in the highest quality format, while reducing the need for vast amounts of storage space for everyday activities.

Live camera viewing

Figure 28-9 is a screen from the ReCam viewing window, showing the video image and its details. Reading across the bottom, the "Capture Rate" is the fps that the

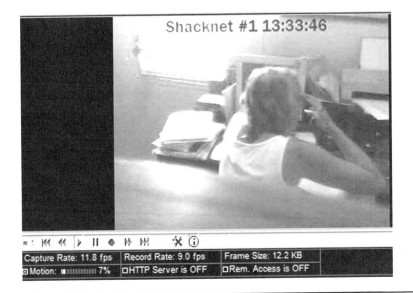

Figure 28-9 ReCam ShackNet screen.

ReCam-equipped computer is receiving from this particular camera, in this case 11.8 fps. This camera is set to record at 9 fps; the recording rate can be different than the capture rate, based on settings in the "Video Compression" screen. "Frame Size" is the size of the stored images, in this case 12.2 kilobits.

Below the first line of information, the "Motion" box is checked, meaning that motion detection is active for this image stream. The strength bars indicate the percentage of motion being detected. Both the "HTTP Server" and "Remote Access" services are turned off.

The icons above the image data provide instantly selectable recording and play-back options. The person monitoring the system can "turn on" and "off" record-ing at his option, as well as reviewing the temporary "live" recorded images.

Multiple cameras can be viewed simultaneously, and up to 32 IP cameras or a combination of cameras and video servers can be viewed, recorded, and controlled using this software package.

Video management software review

While virtually any IP network camera or video server can be accessed and viewed using a standard web browser program, video management software such as ReCam provides many valuable options for video image viewing, management, and storage.

CHAPTER 29

Digital Video Recorders

The Digital Video Recorder (DVR) has become a popular upgrade for existing analog CCTV systems, as well as the preferred method of camera control and recording for many new installations.

The following section will detail some of the common features found in DVRs, with a particular focus on the applications and advantages of network-enabled DVRs. The new breed of video capture technology, which converts a standard desktop PC into a powerful DVR, will also be examined.

DVR connections and functions

The typical DVR will accept a maximum of 9 or 16 analog composite video signals, connected via coaxial cable and BNC connectors. Video signals reaching the DVR are sampled into a digital format, after which one of various methods of video compression is applied to reduce the size of the images for storage. Images are stored onto one or multiple hard drives.

Recording options

DVRs will typically store video images with a maximum number of frames per second. This maximum storage capability must be shared by all connected and recorded cameras. So if a DVR provides a maximum of 480 fps recording capability, and it has 16 cameras connected to it, 30 fps or "real time video" can be recorded, as 16 times 30 equals 480.

Although this would appear to be the proper way to use the DVR's recording capabilities, the massive storage requirements for real-time video generally force the installation company and/or security management to reduce the recorded fps.

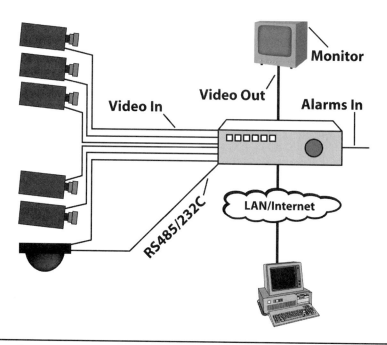

Figure 29-1   DVR connectors.

In our example, if the file size of each image is 12 kilobits, the storage requirement for one minute of video from one camera would be as follows:

(12,000 bits times 30 images per second) times 60 seconds = 21,600,000 bits

or 21 megabits of storage capacity for one minute of real time video. As hard drive storage space is sometimes measured in bytes (one byte equals eight bits), the storage requirement for the above example can also be expressed as 2.7 megabytes (21,600,000 divided by eight).

One advantage of old-school VHS tape recording is enormous bandwidth storage capabilities; DVR systems must be programmed to provide a minimum amount of fps recording so as to maintain a reasonable volume of recorded data, usually measured in days or weeks, on the DVR's storage disk(s).

DVR manufacturers will often use very low fps rates in their literature, as the calculated storage capacity of the DVR's hard drive is greatly expanded if only one or two frames per second are being recorded from each connected camera. One vendor, whose DVR is provided with one 120 GB storage drive, says that its unit will store one week of video from nine cameras, with each camera being recorded at 1.66 frames per second.

The problem with this approach is that a lot can happen in half a second, particularly if people or perpetrators are moving across the field of view. It is quite conceivable that by setting up a DVR to record as little as 1.6 fps per camera, there may be little or no useful video of a fast-moving event or person that can be extracted from the DVR for evidentiary purposes.

Two technologies that are often used in DVR setups to provide higher fps recording for "events" that should be recorded at higher speeds are video motion detection and expanded storage capability.

Video motion detection

The basic concept of video motion detection (VMD) is a measurable change in the light amount in a pixel or array of pixels on an image from one frame to the next. If a selected area of an image has gone from lighter to darker, the video motion detection software "sees" this change as motion, and can initiate a variety of actions based on how it was programmed. Video motion detection can be included in network cameras, or analog CCTV images can have motion detection applied to them by the DVR after the analog video signal has been digitally sampled.

Programming the image of a DVR-connected camera is usually a simple process of selecting a "motion area" or set of pixels whose brightness change will trigger faster fps recording, usually for a short time duration such as 30 seconds or one minute after the motion has been detected. Some DVRs also offer pre-alarm recording, where a programmable number of seconds of video prior to the motion detection event will also be recorded. A DVR is actually a full-blown microprocessor computer, utilizing many of the same chip sets and hardware found in a typical desktop PC. So the "magic" of recording images prior to the occurrence of an event is provided by the DVR, which has been programmed to constantly store in memory (or "buffer") a minimum amount of video images equal to the pre-alarm recording time period selected. If the motion detection or another wired alarm input is triggered, the buffered video images are transferred to hard disk storage, along with the post-alarm images.

A recent innovation in motion detection is the post-event application of motion detection technology. Panasonic is one vendor that supplies a DVR that provides the ability to, for example, place a motion detection "zone" around the image of a car in the parking lot as transmitted by a stationary camera. The DVR will now search for movement around the vehicle, and the person getting in the car and driving away can be viewed.

Panasonic also provides a "Vector" VMD program, which can detect motion based on the direction of movement. For example, perhaps a camera will be located to view the entryway of a building from the side; movement toward the building triggers VMD features, while movement away from the building does not.

Most DVRs can be programmed to continuously record other non-triggered camera images at their usual rate while immediately increasing the fps of the motion-triggered camera.

SECURITY TECHNICIAN'S NOTE: Video motion detection works great with stationary cameras, and not very well at all with moving pan/tilt/zoom CCTV cameras. If a camera has motion detection enabled and it is panned, tilted, or zoomed, the changing image on the screen will trigger the motion detection, increasing fps recording even though the actual movement is detected from the camera's movement and not from any person or thing moving in the field of view.

Wired alarm triggers are also available, providing interfacing capabilities for access control and intrusion detection systems. For example, a relay output from an access control system can be programmed to activate recording or faster recording of a specific camera connected to a DVR. This relay activation could be produced by a person presenting an expired access control card at a specific reader location.

Disk storage options

Simply put, the more gigabits of storage, the more video images can be kept on the DVR. Most DVRs have options for additional hard drive capacity, sometimes using plug-in hard drive modules or connected hard disk drive (HDD) devices. Using complex Redundant Array of Independent Disks (RAID) technology, massive storage of video images up to 12 terabytes (one terabyte equals 1,000 gigabytes) and beyond can be provided. RAID technology uses a number of read/write hard disks that are seen by the network or controlling computer's operating system as a single logical storage and retrieval location. Using "striping," sections of a saved file are spread across the disk array. This provides for quicker retrieval of large files, as all of the read/write hard disks can simultaneously access their portion of the file that has been requested, rather than having the entire file on a single drive. RAID systems can be programmed for various forms of redundancy across the disk arrays so that the catastrophic failure of one hard drive will not cause the loss of stored data.

Evidentiary options

DVRs will generally provide some method of evidentiary video extraction so that the video images of the Bad Guys can be recorded onto a CD/ROM or DVD to be given to the authorities. Some recorders provide an analog RCA or S-Video output that can be recorded onto a VHS tape device. Such outputs can also be utilized to provide long-term digital storage of video images.

Viewing options

DVRs can replace the typical multiplexer, offering multiple screen viewing with "full duplex" capabilities, where a camera's previous images can be viewed while its current images are still being recorded. Different vendors offer various options for viewing while simultaneously recording. Switching functions are also available on some DVR models.

It is important to note that while a DVR may be programmed to only record a few fps of a particular camera, the "live" view of the monitor will be in full motion. On playback, the stored images will present a "slow motion" view of the activity, based on the number of frames per second recorded.

P/T/Z control

Most DVRs offer some form of pan/tilt/zoom (p/t/z) control for dome cameras. This control is often provided by a separate RS-232 or RS-422 output from the DVR. While some DVR manufacturers provide the p/t/z control protocols to control cameras from a variety of vendors, there are others whose DVR will only control p/t/z cameras manufactured by that DVR manufacturer.

Networked DVR benefits

While all of the features mentioned above are available in standard DVRs, those that can be connected to a LAN and/or the Internet provide a tremendous upgrade in control and features. Putting the DVR on the network allows authorized users to use their computers to view and control the images and cameras as if they were pushing the buttons on the face of the device. Monitoring of a building's cameras can be transferred to another location during off-hours, reducing guard costs. Email notifications can be sent out when there is an alarm or motion is detected, or if there's a problem with the DVR itself. Multiple users can view the images at the same time from locations all around the world. Image files can be copied from the DVR's storage onto the distant computer, again by authorized users.

> **SECURITY TECHNICIAN'S NOTE:** Network DVRs will utilize either an embedded Windows or Linux operating system. Many network IT managers do not want to place embedded Windows devices on their network because of security concerns. Security company personnel should know what operating system their devices use. If installing Windows-based equipment, security technicians should check with the manufacturer to see how periodic "security patches" issued by Microsoft can be installed into the specific device.

Connection issues for network DVRs

Network DVRs are connected onto the Ethernet LAN in the same manner as an IP camera or video server. Installation technicians will need to input the proper type of IP address and access port numbers and clear the firewall paths from the DVR to the Internet connection, provided that access to the DVR from the Internet is desired.

DVRs and hacker security

Because of their limited feature set and software options, most DVRs are considered to have "proprietary" software and programming. This is touted as a plus by DVR vendors, who say that because their products are not based on standard PC technology they are less prone to hacker attacks. However, as explained previously, most hacker attacks are borne by infected email messages. If a network-connected device doesn't have an email address, it is much less susceptible to hacker entry. And although DVRs may be able to send out email messages, they generally don't have the capability to receive email.

Desktop DVRs—video capture systems

As a DVR is really a specialized version of a PC-type computer, it comes as no surprise that enterprising vendors have developed technology that adds DVR capabilities to standard desktop PCs.

Converting a desktop PC into a DVR starts with a "video capture card," which fits into one of the open slot bays within the computer pedestal. This card typically provides BNC connections for four, eight, or sixteen analog camera inputs. The

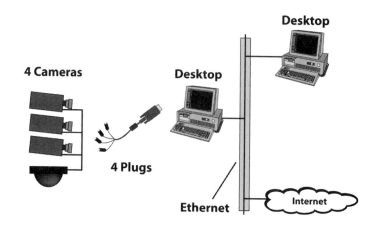

Figure 29-2 Desktop DVR.

capture card performs the digital sampling, converting the composite analog video signals into digital bit streams, which are passed to software in the computer for compression, storage, and display. Desktop DVR systems can provide many of the features of stand-alone DVRs detailed above.

The advantage of desktop DVRs is primarily cost, as the capture card and software can either be installed in an existing PC, or a new low-cost computer can be purchased and equipped. Using standard computer hard drives for additional storage can greatly reduce the overall cost of providing DVR benefits to the end user.

The disadvantages to desktop DVRs can be summed up in a single word: security. Because it is a desktop PC, with a keyboard and mouse, authorized or unauthorized users have the opportunity to "hack" at the system, potentially disabling critical recording functions. If the desktop DVR is connected to the Internet and receives email messages, viruses and other hacker programs have the potential to enter the computer. However, adequate defenses from such potential attacks can be readily performed, as is detailed in another section of this guide. A summary of defense options includes dedicating the desktop PC to DVR functions only, eliminating Internet access for users from that PC, removing CD-ROM and floppy disk portable media drives, and installing anti-virus software, which should be used regularly to perform virus scanning and regularly updated.

Desktop DVR products offer a viable alternative to stand-alone DVR systems. Using off-the-shelf components greatly reduces the cost, and having the system reside on a "regular" computer may reduce training time for users who are already familiar with using a typical computer software program.

Benefits of DVRs

Whether stand-alone or desktop (i.e. PC-based), DVRs provide the digital link between analog cameras and networking benefits. Perfectly adaptable to existing installations, DVR products provide superior recording, camera control, and switching options, when compared to older technology equipment.

Although the DVR is an interim technology between tape-based VHS recording systems and networked digital storage, the preponderance of previously installed analog camera CCTV systems signals a long future for DVRs in security applications, because systems that use networked digital storage are not compatible with analog cameras.

Summary

There are a wide variety of DVR technologies available to provide digital storage of connected analog cameras as well as local and remote network connection capability for authorized users. While some DVRs are in a single box, others may be video capture boards installed into a desktop PC. Some DVRs use an embedded Windows operating system, while others use some variation of Linux. Security dealers should shop around, as there are many different DVR features and price points.

CHAPTER 30

IP Alarm Transmitters

As networks proliferate, security equipment manufacturers have developed IP alarm transmitters. These devices interface with burglar and fire alarm control panels, transmitting alarm, trouble, and status reports over the LAN and/or Internet.

This section of the guide provides a general discussion of IP alarm transmitters, their installation, and application issues.

Why transmit alarm signals over networks?

One of the primary benefits of IP-networked alarm signals is that signals can be received and processed from remote locations by sending the information over the Internet. By concentrating the monitoring of a number of buildings into a single location, client costs for guards are reduced. Also, IP alarm signals can be monitored at multiple locations, providing an alternative in the event that the primary monitoring station has been disabled.

Another key advantage of IP transmitters is higher security, as the devices can be "polled" by the central station every few minutes for functionality. IP transmitters are approved for use for fire alarm transmission, UL Standard 864, as well as UL Standard 1610 Intrusion Systems, provided that they are properly installed, programmed, and monitored. Using IP alarm transmitters also eliminates issues of trying to connect and communicate alarm signal transmissions over VoIP connections.

At the central station end, supervisory and alarm messages from IP alarm transmitters can be received by a properly equipped alarm receiver, which usually requires a specific "line card" connected to a broadband Internet connection. As processed by the central station's alarm monitoring software, there is no difference between the receipt of IP alarm messages and standard digital communications. IP alarm transmitters must be programmed to generate "polling" messages that confirm remote transmitters' functionality with the central station on a selectable time basis. UL installations will require a specific maximum time between polling messages.

Pick a winner

There are a number of issues that alarm installation companies must address to successfully install an IP alarm transmitter at a particular location. The first is product selection . . . which IP transmitter should be installed?

The alarm industry has become dependent on the telco-connected digital communicator, which can be programmed to provide a wealth of different types of alarm signals, such as user IDs, open/close, system troubles, and specific zone or device alarm activations. An IP alarm transmitter can provide the same depth of information to the central station, provided that it is properly matched for the control/communicator at the client's location.

The selection of the proper IP transmitter to use is very often dependent on what specific burglar alarm panel is being connected. Major manufacturers such as DMP, Honeywell, Bosch, and DSC provide specific IP transmitter interfaces for their popular products, allowing a relatively simple connection to the panel and a full range of reporting options.

If no "matching" IP transmitter is available for a particular panel, the panel can be replaced, or there are other options. An installing company can use a "Dialer Capture Module," such as one available from Bosch. This device connects to the digital dialer outputs of the panel and emulates a digital receiver. When an alarm signal is transmitted, the DCM accepts the digital output, transmits the signal over the IP network to the central station, and simulates the alarm receiver's acknowledge messages to the alarm control after the successful transmission of the IP alarm signal.

If the client's control panel provides auxiliary voltage triggers or relay outputs, these may be interfaced into IP alarm transmitters that accept those types of inputs. The TL-250 IP transmitter from DSC provides a number of possibilities for connection to existing panels, as well as mating to DSC controls. If the client's existing control panel only has simple pulse/steady bell outputs, those can be connected and transmitted as separate messages. Controls with separate zone or status outputs also can be interfaced with up to 4 separate reports (or 12 if the TL-250 is equipped with an expander board). This product will also function as a stand-alone IP alarm/status transmitter, allowing the connection of environmental sensors, such as temperature monitors, or any other sensor that provides an output.

Receiver compatibility

Current IP transmitter products will only communicate with receivers manufactured by the same vendor. So the type of central station receiver that will process the signals may well determine selection of a transmitter vendor. If, for example, the central station uses a Bosch/Radionics receiver, Bosch IP alarm transmitters must be used.

Honeywell provides IP alarm transmission capability to participating central stations through the AlarmNet network, which also provides long-range radio and cellular alarm capability. Customers don't need a radio transmitter at the premises; instead, using a broadband link, they connect to a network access point on the Alarmnet network.

Losing the signals

If the IP alarm transmitter is connected to the alarm panel via relay outputs or alarm triggers, the variety and amount of individual alarm messages available for transmission to the central station will be greatly limited, as compared to standard digital alarm signals. Detailed signals such as individual zone alarms, user-specific open/close reports, and other such expansive alarm and supervisory signals may be impossible to pass to the IP alarm transmitter.

Pay to play

As when other technologies have been introduced into the electronic security industry, the first generations of IP alarm transmitters are not inexpensive. Typical costs are in the range of $250 or more per unit, although manufacturers will soon be releasing devices at substantially reduced costs.

Installing an IP alarm transmitter

There are several issues to consider when installing an IP alarm transmitter. The first is the data connection to the network. A network connection, typically a UTP Cat 5e or higher quality, must be available to connect the IP transmitter, located in or near the alarm control panel, to an available port in the IP network. This may be an otherwise unused RJ-45 jack on the wall, or new UTP cable may need to be run to bring the network to the alarm control. It's possible that the alarm panel may be installed within the main cross-connection or horizontal cross-connection room, making connection to the network a simple process. See Chapter 14 for more information on cabling options.

General programming

Once the IP transmitter is in place, it must be programmed to the proper central station ID and other selectable features. Programming methods will vary with different manufacturers as well as the particular central station receiver being used. Although a particular IP transmitter may be programmable over the network, use of a laptop computer will provide the technician with a quick connection to the transmitter, and easy viewing of the transmitter's indicating LED array. Some vendors' products are programmed with a specific device, such as the Honeywell/AlarmNet 7845I, which uses the 7720P programmer, which also functions as the programmer for Honeywell's AlarmNet radio and cellular products. Programming through a "mated" control panel's keypad is available from vendors such as DSC.

Along with programming the IP alarm transmitter, the control panel outputs may need to be "turned on" via a keypad or programming selection, and the central station receiver must be programmed to receive alarm signals and supervisory messages from the field device.

IP addressing

As described in Chapters 11 and 12, network devices must each have a unique IP address to allow them to communicate with other devices on the local network, and access the Internet.

Testing of alarm transmissions

As with any type of alarm signal transmission technology, complete and thorough testing of all alarm signals to the central station should be performed by the installer before completing the installation and leaving the jobsite.

Digital dialer backup

While IP alarm transmitters provide much higher security than telco-connected digital communicators, connecting a telephone line to the digital communicator outputs of a control panel can provide dual and/or backup reporting. This would be desirable in the event of IP transmitter or network failure. Client or telephone network use of VoIP (Voice over Internet Protocol) may negate this feature, as some digital formats will not pass through this digitized voice protocol, and a VoIP connection often becomes inoperable when the computer network goes down.

> **SECURITY TECHNICIAN'S NOTE:** While individual network components can fail at a client's location, Internet services are vulnerable to attack by outside forces. In a Distributed Denial of Service (DDoS) attack, hundreds or thousands of remotely controlled Internet computers will bombard a specific host on the Internet with millions of packets, causing the target device to be unavailable to its regular users. It is only a matter of time before such an attack is directed at the IP address of a central station's Internet alarm receiver.
>
> It is very important to provide some method of redundant alarm communication along with an IP alarm transmitter for high-security applications.

Loss of downloading capability

The vital technology of remote downloading of control panel functions can be affected by the installation of IP alarm transmitters. If the IP network is the only communication connection to a particular panel, telco-based downloading will not be available for that system. Some vendors such as DSC provide for remote downloading capability over the Internet by using a specific software set in the downloading computer.

Transmission options

Some transmitters can be programmed to transmit alarm signals to multiple IP-addressed receivers, providing redundant or back-up transmissions. This is a powerful feature, as critical alarm messages can be transmitted to a traditional central station, while supervisory or less important transmissions (along with critical burg/fire messages) can be received and responded to by the client's own monitoring facility.

Signal encryption and security issues

With the ready availability of literally millions of Internet connections, alarm equipment vendors have built sophisticated signal encryption into IP alarm transmitters and receivers. Signal encryption defends against a "rogue" transmitter being used to emulate "I'm OK, no alarm signals here" polling signals emanating from an IP alarm transmitter at a target installation. Such encryption programs include the use of "keys," which are held at both the receiver and transmitter. When an encrypted communication is received, the key is used to convert the garbled signal into its original state. Typical encryption programs used are "Blowfish," which uses a 1024-bit format, and 128- and 256-bit AES.

Programming of the encryption format to be used will generally be dictated by the products selected and the encryption program used by the corresponding alarm receiver at the central station.

Future/now

IP alarm transmitters provide much higher security of alarm signaling, and will become more popular as IP networks grow and the traditional telco-based alarm monitoring paths mutate into hybrid VoIP/analog networks.

How should an alarm installing company prepare itself for installing IP alarm transmitters? Here are two recommendations:

1. Transmitter Selection—Determine which IP transmitter(s) will best function with the existing panel population and central station receiver

2. Panel Selection—Consider which alarm control panels should be installed on new jobs, so that the migration path to IP transmission is a simple process

Practice, practice

Installation of IP alarm transmitters can be tricky, as new programming techniques and communication methods are needed. An intelligent approach for alarm companies is to practice at the office by bench-test connecting IP alarm transmitters to the panels the company typically uses in the field. Once the glitches have been surmounted, field installations can proceed without undue difficulties.

Summary

Burglary, fire, holdup, and other supervisory alarm messages can be transmitted to a central station via IP-enabled alarm transmitters. These devices can be connected to most current alarm controls and can be configured to provide various grades of UL-approved alarm signaling. IP alarm transmitters must be carefully selected based on the panel to be monitored and the type(s) of alarm signals that need to be transmitted. Traditional digital communicators are often used to provide a redundant communications path to the central station.

CHAPTER 31

VoIP and Alarm Communications

Like a black thundercloud approaching, the growing popularity of "Voice over Internet Protocol" (VoIP) portends great problems for the electronic security dealer with digitally monitored accounts. VoIP is a nascent technology that is growing in popularity with homeowners and businesses wishing to reduce their costs for local and long distance voice telephone calls. If alarm monitoring subscribers convert their telephone service to this IP network technology, they may well disconnect or disrupt their alarm panel's connection to the central station. And VoIP technology is supplanting switched communications between telco central offices, possibly preventing the transmission of certain popular digital communicator formats.

VoIP is a "clear and present danger" to traditional alarm communications, and may signal the future demise of the industry's communication workhorse, the digital communicator. The potential and growth of VoIP indicates that the phone line may be going away as the primary platform for alarm monitoring communications.

How VoIP works

The technology of VoIP emulates the functions of a standard switched telephone line, but uses the Internet or other IP network to provide the pathway for telephone calls. An IP-addressed interface device is connected to the Internet via a cable modem, DSL adapter, or other ISP link, and provides connections for standard telephone instruments. Once activated and programmed, the VoIP adapter provides a dial tone to the telephone instrument. When a call is placed, the analog sounds from the telephone are sampled and converted into IP data packets and transmitted over the Internet, where they are converted back into analog at the receiving end.

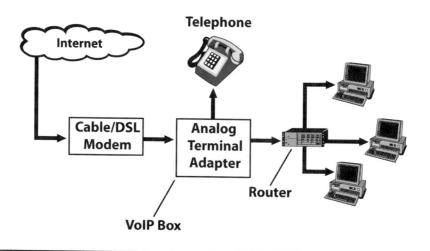

Figure 31-1 VoIP connections.

This technology is also being implemented by local and long distance telephone companies to carry telephone traffic between offices. So it is quite possible that if alarm subscribers' central stations are located across state lines, at some point in the near future their digital alarm signals may be traveling over VoIP to reach the alarm receiver.

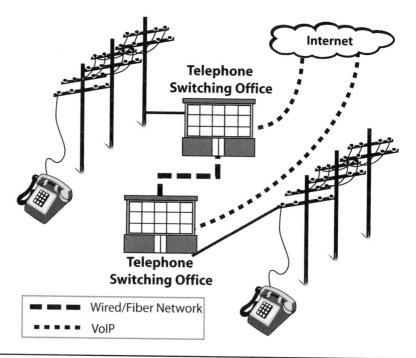

Figure 31-2 VoIP in the telephone network.

Why your customers may switch to VoIP

The primary attraction for alarm subscribers to convert from traditional telephone companies to VoIP is lower cost. Customers receive flat rate billing for local and long distance calls at substantial savings. In most cases subscribers can retain their existing telephone number.

Some VoIP providers encourage subscribers to keep their traditional telephone service for emergency use because, unlike traditional telephone connections, VoIP telephones and DSL or cable modems will not operate during power outages unless separate backup power is provided. Also, some telephone companies will not provide DSL service unless customers also purchase traditional voice service.

Another benefit of VoIP is that subscribers can get a phone number outside their own area code. This offers commercial clients the option of having local phone numbers in Dallas, Los Angeles, and New York while the calls are actually being answered in Columbus, Ohio, for example. Many middle- to large-size companies are using VoIP to connect their markets to one call center.

VoIP and service quality

As was experienced with the initial phases of the deployment of cellular telephones, new technologies often go through an early period of trials and tribulations as manufacturers, installers, and users learn to best utilize the benefits while minimizing the effects of problems.

VoIP is going through that initial phase now, with some problems with service availability, sound quality, and other issues reported by users. Telecom providers are also working together to standardize VoIP protocols and communications to provide seamless communications between their networks.

Installation problems with VoIP

Let's look at the example of a typical residential security system that is monitored via a digital communicator connected to an incoming switched telephone line. This customer uses a cable modem to provide a broadband Internet connection.

If this subscriber chooses to connect VoIP, he will install it himself after receiving a VoIP adapter and software from a provider such as Vonage.

As long as the subscriber retains the existing telephone connection to the alarm panel, security communications would continue uninterrupted. But remember that the primary reason for switching to VoIP is saving money, so it is quite possible that the subscriber will cancel his traditional telephone service and rely on the cable modem to carry telephone calls and Internet traffic. If this occurs, the alarm system will be unable to communicate.

Unless the subscriber calls the alarm company to ask about connecting his alarm to a VoIP connection and/or the subscriber's alarm equipment is programmed to transmit a periodic test message, there will likely be *no indication that the alarm system's transmission capability has been disabled* until there is an actual alarm or the client attempts to test his system into the central station.

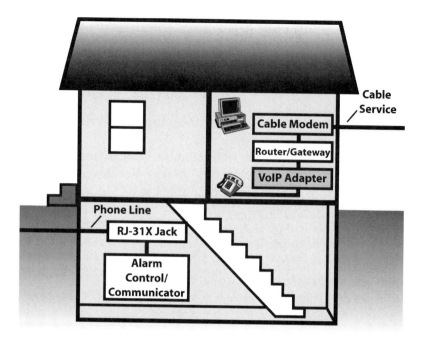

Figure 31-3 Home office VoIP.

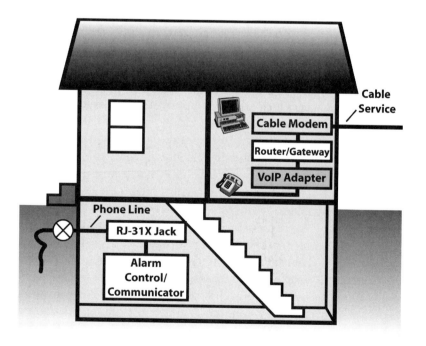

Figure 31-4 Digital dialer disconnection.

Alarm transmission problems with VoIP

Another concern pertains to whether a particular alarm panel/communicator's digital format will successfully reach the central station receiver over VoIP.

Alarm equipment manufacturers and industry associations have been receiving reports of the incompatibility of certain dialer formats with VoIP. Preliminary testing by CANASA and some manufacturers indicates that "pulsed" formats such as 3/1, 4/2, SIA, and Radionics modem formats may work, while the DTMF formats such as Contact ID may not.

Testing for format compatibility with VoIP poses many variables, making it difficult to say definitely whether one will or won't function. Different end users may use VoIP interfaces from different suppliers, just as IP network switches and routing equipment can come from a variety of vendors. Although alarm industry communicators theoretically should work with these devices, problems of network "latency" can frustrate or defeat the transmission of alarm signals. "Latency" is the concept of the different lengths of time data packets take to travel from one network device to another. When an ISP is in heavy use, data packets will take a longer amount of time to reach their destination, which likely contributes to the instances of problems when transmitting DTMF formats. Many industry experts believe that this time variance issue would make the successful transmission of digital communicators over VoIP a "hit or miss" proposition at best. Additionally, VoIP providers use a wide range of codecs to compress analog voice signals for digital transmission and some of these, particularly those requiring the least bandwidth, also have caused problems with alarm transmission in tests conducted by several different VoIP equipment vendors, as well as alarm industry tests.

These potential variables leave alarm companies with uncomfortable possibilities when connecting digital communicators to VoIP. If subscribers change their VoIP service, ISP, or if the Internet backbone components are changed or rerouted, the communicator may fail to function. It is also important to realize that after the VoIP interface converts the digital communicator's output into IP data packets, those packets can take various routes through the Internet to reach their final destination. Data packets can and will route through different network segments during a single transmission, so part of the information may reach the central station in usable form while other parts may not, resulting in a failed communication. And although digital communicators are usually programmed to retry communications in the event of failure, there is no guarantee that such retries will be successful. Even if the repeated communication does reach the monitoring station, there would be an increased delay in dispatching time.

Connection of digital communicators to VoIP

It's important to note that many VoIP providers do not recommend the connection of alarm communicators to VoIP lines. Due to the problems of power backup and communicator formats, alarm communicator manufacturers in general are also not encouraging such connections.

One problem confronting the installer when the subscriber changes to VoIP is establishing a "phone line" connection between the alarm panel/communicator and the VoIP interface. As is seen in the previous illustrations, often the alarm communicator is located in a basement, closet, or storeroom far removed from the VoIP interface, which will likely be in the subscriber's office. To address this, new cable would need to be installed between the panel and the VoIP interface location, and proper termination to the new VoIP "telephone" line would be needed to provide for line seizure and alarm communications.

With regard to power backup, a CANASA report titled "Voice over Internet Protocol" suggests that some 12-volt DC VoIP adapters can be backed up from a power supply with a standby battery. This arrangement could be integrated into the alarm system for power supervision. The DSL adapter or cable modem might need a separate UPS (Uninterruptable Power Supply) to provide backup power.

Alternatives to digital communications

If a subscriber has converted his telephone service to VoIP or if the telco central office has switched to VoIP for interoffice traffic, disabling digital communications, how can alarm signals be restored and maintained? There are a number of alternative communication technologies available to the industry, including wireless transmitters such as long-range radio and cellular. These alternatives can work well, provided that there is sufficient coverage of the wireless network chosen, and that the alarm company's central station is equipped to receive such signaling.

Another method of alarm monitoring turns the subscriber's Internet connection into the primary path for alarm signaling. DMP (Digital Monitoring Products), Bosch, DSC, ADEMCO/Alarmnet, and others manufacture a variety of alarm control/communicators and "slave" devices that transmit alarm information directly over the Internet. By connecting the panel or communicator to an Ethernet network that is routed to the Internet as described in Chapter 30, packetized alarm signals can travel to a properly equipped central station anywhere in the world.

Death of the digital dialer

Just as the electronic security industry has witnessed the obsolescence of other communication technologies such as tape dialers and direct wires, many industry experts are convinced that the death of the digital communicator is right around the corner. Alarm equipment manufacturers are either currently delivering or quickly preparing for the anticipated growth of IP transmitter usage.

How dealers can protect their business

There are three strategies that installing companies should employ to protect their monitored accounts and the revenue associated with them.

1. Notify subscribers—written notices, letters, telephone calls, and emails can all be used to warn subscribers of the potential danger to their alarm system if they select VoIP services.

2. Reprogram panels to VoIP-friendly formats and in anticipation of the growth of VoIP, use VoIP-friendly formats for all new installations.

3. Program all accounts for periodic test messages—if subscribers disconnect their alarm panels by choosing VoIP over a cable modem and then canceling their traditional phone service, periodic test messages are the only way that an installing company can be notified of the communicator's disconnection *BEFORE* an important alarm activation fails to reach the central station. These test messages are also critical to protecting the integrity of digital alarm monitoring in the event of VoIP technology being introduced into the "back channels" of the telecommunications networks.

Although a test message option has been available in alarm communicators for many years, a great majority of digitally monitored accounts are not enabled for this vital function. Contract central stations have indicated that the percentage of accounts programmed for this feature, outside of those with open/close reporting, is as low as 3%, leaving well over 90% of such accounts potentially unmonitored in the event of VoIP implementation, either at the clients' location or within the telecom network. One reason for this is economic, as contract central stations typically charge an additional fee for test message processing.

> **SECURITY TECHNICIAN'S NOTE:** If the cable company or other provider is making a major push into a security company's market area, security dealers should test their common control panels and digital communication formats on the typical VoIP adapters provided by the VoIP vendor.

What's the good news?

There is a way for alarm companies to benefit from customers going to VoIP. Clients with VoIP inherently have a broadband Internet connection with either a cable modem, T1 line, or DSL interface, and are prime candidates for the installation of wired or Wi-Fi network cameras. These devices can provide alarm company clients with the ability to view video images (and with certain products, to receive audio as well) of their home or business from any Internet-connected computer. Such subscribers may also be upgraded from standard digital alarm communications to IP network transmitters, which can provide regular polling of the transmitter for functionality by the central station to verify its functionality.

It's somebody else's problem

The telecommunications networks, which the industry relies on to provide the conduit for alarm monitoring and the revenue it provides, most likely will not modify their VoIP services to accommodate alarm industry digital dialer communications.

Alarm equipment manufacturers are delivering alternate devices for alarm transmission, and installing companies will have to shoulder the burden of converting paying subscribers onto IP or other alarm communications technologies or risk failed communications, lawsuits, and reduced revenues from subscriber cancellations.

Summary

Many security system customers are converting their traditional telephone service to VoIP, which transmits voice telephone calls over the Internet. Digital alarm communicators connected to VoIP telephone adapters may or may not transmit successfully to the central station when activated. Technicians may need to change the communications format in an alarm transmitter to enable communications over VoIP. All accounts should be programmed to provide periodic "I'm OK" test messaging to the central station, as clients may convert to VoIP at any time and it is likely that they may not notify their electronic security company of the change.

CHAPTER 32

Tools of the Trade

Although just plugging in devices and programming them from a local PC can successfully perform many network installations, specialized tools and testers can make installations and troubleshooting much easier for security technicians.

The following is a list of recommended equipment and tools for security networking. We'll look at each, along with an explanation of its use.

Ethernet tools

1. Laptop Computer—This computer should be equipped with an Ethernet port, Wi-Fi communications, and Windows software. This computer can be used to program networked devices and troubleshoot communications problems. A spare battery pack, charged up prior to visiting the jobsite, is a smart idea.

2. Ethernet and Cabling Tester Set—This will test for 10/100 Ethernet communications on UTP links and whether the pin connections on an RJ-45 socket are correct. These testers are low cost—less than $150US. The "TVR 10/100" manufactured by the Byte Brothers is a good example of this type of tester.

3. Crimping Tool Set—Technicians will need to crimp male RJ-45 connectors onto newly installed UTP cables.

4. "110/66" Punch Tool—Used to terminate UTP conductors onto Cat 5 jacks. Cheap ones are hard to work with and don't last very long. Purchase a quality tool.

5. UTP jumpers—Technicians should carry standard and crossover UTP jumpers for programming and testing of network devices.

6. A small hub or switch for connecting multiple devices.

Wi-Fi tools

1. Laptop Computer—Use a Wi-Fi-enabled laptop to test for Wi-Fi coverage areas. Program the laptop to communicate with the Wi-Fi access point or router, and test for sufficient signal strength at the location where a Wi-Fi camera or other device is to be installed.

Product-dependent tools

1. Cable(s) and/or interface card(s) needed for setup, configuration, or normal communications.

2. Software—may require installation or registration keys or codes to enable features.

CHAPTER 33

Testing and Troubleshooting

While the physical connection of networked electronic security devices to an Ethernet or Wi-Fi network can be accomplished by powering the device and plugging in a UTP jumper, programming and addressing can be complicated. After powering and programming a device, using a standardized testing procedure will help technicians verify and achieve the proper communications.

Logical networks

For successful troubleshooting, it's important to understand that computer networks function in a logical fashion, with hardware and software performing their functions in a "yes-or-no," "on-or-off" manner. Computers don't suddenly decide to do things in a different way than previously; only software or hardware configuration changes will cause different actions in a computer or the network.

If a technician attempts to communicate with a networked device by inputting the device's IP address and the communication doesn't work, repeated mashing of the "Enter" key will not improve or change the situation. Network devices do what they're told. If a device isn't functioning properly, a change in the programming of the device itself, or some device in the path, is needed to achieve functionality.

Communications testing sequence

As an example, the following sequence details the communications testing procedures for an Ethernet network camera, DVR, or video server. Testing procedures for other types of security devices will be similar, with the important addition of final functionality testing of the specific devices.

After powering, programming, and connecting the network video device, the three steps to testing are:

1. Direct Laptop Testing

2. LAN Communications

3. WAN/Internet Communications

To avoid confusion, these tests should be performed in the order listed above.

Direct laptop testing

Directly testing the device with a laptop ensures that the device is properly programmed for its IP address and can communicate with an Ethernet-enabled PC, in this case the laptop.

The steps for direct laptop testing are:

1. Set the laptop's IP address to be on the same network as the device to be tested (Chapter 11).

2. Enable the laptop's Ethernet NIC.

3. Connect the laptop and the device with a crossover UTP jumper.

4. Turn off any software firewall programs currently running in the laptop.

5. Open a web browser, such as Internet Explorer, enter the IP address of the device, and press "Enter."

6. The username/password screen of the device should appear on the laptop.

7. Sign onto the device.

8. Test for video image transmission, i.e. does the video image appear on the laptop.

9. If OK, close out of communications, remove the crossover connection, and reconnect the device to the LAN cabling, i.e., the hub, switch, or router where it is supposed to be connected.

If the device will not communicate with a directly connected laptop, the cause may be one (or more) of the following:

1. The technician forgot to shut off the software firewall in the laptop.

2. The laptop and device are not programmed to the same LAN network address range.

3. The UTP jumper is not a crossover cable, or is defective.

Direct laptop testing is the first step in confirming the programming settings of a networked security device. If the device won't communicate with the laptop, it won't communicate with the network.

LAN communication testing

Once the laptop test is completed and the target device has been reconnected to the LAN cabling, communications can be checked between other computers on the LAN and the new device.

The steps for LAN communication testing are:

1. At a PC connected to the LAN, turn off any software firewall programs currently running; open a web browser, such as Internet Explorer. Enter the IP address of the device to be tested and press "Enter."

2. The username/password screen of the device should appear on the PC's screen.

3. Sign onto the device.

4. Test for video image transmission, i.e. does the video image appear on the PC's screen.

5. If OK, close out.

If the networked device will communicate during the direct laptop test but doesn't from a LAN PC, here are the possible problems:

1. The PC's software firewall is blocking the communication. Turn it off.

2. Addressing error—the IP address in the device isn't in the same range as the LAN PC's. Use the IPCONFIG command (Chapter 11) to check the LAN IP address of the PC. Reprogram the security device or PC IP address if necessary.

3. The LAN PC is not physically on the same network as the device being tested. Are there two (or more) separate networks running at the client's location?

4. The security device is not programmed to the correct "default gateway."

LAN communication testing ensures that local PCs can reach the network security device.

WAN/Internet communications

Being able to view and control networked security devices over the Internet or WAN is the primary reason for their use. This final test confirms that the device is accessible from the outside network. This test should only be performed after the direct laptop and LAN tests have been successful.

The steps for WAN/Internet communications are:

1. Go to a LAN PC, confirm that it has a functional Internet connection by opening a web browser and accessing any valid web site, such as www.slaytonsolutionsltd.com. If OK, proceed to the next step.

2. Using www.whatismyip.com, confirm the current public IP address of the network. Or ask the IT department for the network's public IP address.

3. Using the web browser program, input the public IP address of the network adapter, along with the specific port number programmed into the device being tested. Remember to use the colon, as in this example:

> 12.34.56.231:85. (If the network video device is using a static public IP address, input that address into the browser.)

4. The username/password screen of the device should appear on the PC. It will take a second or two longer than the LAN test; however, if it takes more than 5 seconds or the "Error" page appears, the communications test has failed.

5. If the username/password appears, sign onto the device.

6. Test for passage of video images and/or functionality of the device.

7. If OK, close out.

Note that this test to confirm Internet accessibility of the networked device can be performed through the same ISP connection to which the device is connected. The upstream and downstream communications will flow through the same network connection simultaneously.

If the device cannot be reached over the Internet using this testing procedure, here are some possible causes and remedies:

1. The software firewall in the PC is blocking the communication.

2. A network hardware or software firewall, if present, has not been programmed to allow communication to the network security device.

3. The DSL or cable adapter has not been programmed to allow the security device to "host" communications through the adapter (Chapter 22).

4. The gateway router has not been programmed properly to provide NAT (Chapter 23) for the security device.

5. The wrong public IP address and/or device port number is being used.

Consulting with the IT department may solve these communication problems. Be prepared to discuss with them what is and is not working.

> **SECURITY TECHNICIAN'S NOTE:** A client's Internet upstream and downstream bandwidth can be tested using a variety of no-cost web pages. One example is www.speakeasy.net. These tests can be run on a PC connected to the same Internet connection to which the network video security device is being attached, using Internet Explorer.

Problems with video images

Even if the testing sequences detailed above have been performed successfully, network video devices may require additional testing to ensure their proper performance.

Simply put, being able to reach the username/password screen of a network video device from the Internet is great news for the technician, but there may be other issues that can cause slow video image transmission through the Internet, or no video images at all.

The following explanation assumes that the three tests detailed above have been performed successfully.

No video

If the target device can be accessed over the Internet with its username and password but no video images appear, here are a few possible causes and their solutions:

1. Viewing Software—Some network video products require specific viewing software be either installed or "turned on" in the viewer's PC for images to appear. Some network cameras use "ActiveX" or "Java" programs, which can be selected in the options of the web browser software. Network video products using MPEG compression may have their images viewed by installing specific viewing software that is provided with the device, or downloading same from the vendor's web site.

2. Firewall Software—Some firewall programs may not block the username/password screen but will block the video signals. Adjustment of the firewall program may fix this problem.

3. Dual Ports—Some network video devices use two separate ports, one for control and command and the other for video image streaming. If the video device uses two ports, both must be cleared through the gateway router, firewall(s), and ISP adapters to allow video transmission.

Slow video

In some cases the video device is accessible over the Internet and images can be viewed, but the frames per second rate is slow or unacceptable. Here are some probable causes and solutions:

1. Connection bandwidth limitations—An ISP connection will only provide a certain amount of upstream bandwidth (Chapter 10), which may only allow a certain number of frames per second to pass through the adapter to the Internet. Applying bandwidth controls (Chapter 18) to reduce the size of files will likely increase the number of frames per second that can be remotely viewed. Or the client can upgrade to a larger-bandwidth connection.

2. Proxy Servers—These devices use NAT (Chapter 9) to mask the addresses of local computers connected to them. In this process, each data packet going out over the Internet has its originating IP address changed to that of the proxy server. The proxy server also logs all incoming and outgoing

data packets. These processes take time, which will thereby slow the transmission of image frames. The images transmitted by network video devices are large files, which means many, many data packets that must be processed by the proxy server. Consult with the IT manager to discuss how the video transmission device may be routed around the proxy server, or if the proxy server can be programmed to allow faster passage of the image file packets.

3. Other Users—It is likely that the network security video device is not the only network device simultaneously communicating with the Internet. Other network users need bandwidth too, and if their data transmissions increase, the frames per second of the video transmission will be slowed.

4. Double Trouble—The various potential problems listed above can occur at BOTH ENDS of the communication link simultaneously.

While the proxy server problem can potentially be quickly resolved, the other issues listed may not be so easy to remedy. If the video transmission rate is unacceptable, the technician should try tweaking the scaling, compression ratio, and frames per second settings on the network video device to achieve an acceptable remote viewing image quality and speed.

Summary

Technicians should understand the logical sequence involved in testing or troubleshooting a network-enabled device. Such testing should start with a directly connected laptop to confirm the IP address settings in the device in question. Once the network addressing is confirmed, the next step is to test the device over the LAN to which it is connected. If LAN connectivity cannot be achieved, a network device will not be able to be accessed over the Internet.

APPENDIX A

Common Networking Terms Glossary

192.168.X.X

Commonly used IP address range for LAN networks. This address range, along with a few others, is assigned for private Class C networks by the Internet Assigned Numbers Authority. 192.168.x.x has been assigned for private network use, giving the availability of 256 Class C networks.

568B

Standardized connection format for male and female RJ-45 UTP connection devices. Pin connections are as follows:

1. Orange
2. W/Orange
3. W/Green
4. W/Blue
5. Blue
6. Green
7. Brown
8. W/Brown

802.11b, g, and a

IEEE protocol standards for wireless computer networks ("Wi-Fi"). While 802.11b is in widespread use, 802.11g is gaining in popularity due to its higher bandwidth capability.

Adapter

Connect networks to the Internet. Typical adapters connect DSL and cable Internet services to networks in homes and businesses.

Ad Hoc

Peer-to-Peer (computer-to-device) network. Used in Wi-Fi to connect network devices to each other without using a Wi-Fi access point or router.

Bandwidth

How much data can pass from point to point in a cable or other medium. Bandwidth is measured in megahertz (MHz). Typical UTP will have the capacity of 100–350 MHz. Bandwidth can be reduced by poor cable and connector installation.

Bit

Compression of words "Binary Digit." A bit is either a "1" or a "0." Used in the measurements of data throughput in a network, segment, cable, or fiber. Such measurements are defined as "xxx -bits per second," where "xxx" is

Kilo = 1000 (one thousand)

Mega = 1,000,000 (one million)

Giga = 1,000,000,000 (one billion)

Bridge

Sets of network devices that connect two different LAN network segments together. Bridges use MAC addresses to limit their communications only to each other. Bridges keep local traffic from reaching other LAN segments, and pass packets destined for other LAN segments.

Broadband

Always-on Internet connection in a home or business. Provided by DSL (telephone line), cable, or satellite. Can be purchased with larger bandwidth capabilities if desired or required.

Byte

A unit of data measurement. A byte typically consists of a string of eight bits, representing one character, such as "A."

Cable Modem

Properly termed a "cable adapter." Always-on adapter that provides two-way communications to the Internet via otherwise unused bandwidth from the local cable TV company's coaxial cables.

Classes of Network IP Addresses

Class A—Large networks—typical address—AAA.xxx.xxx.xxx
Class B—Medium WAN networks—BBB.BBB.xxx.xxx
Class C—Small/Private LAN networks—CCC.CCC.CCC.xxx

(See also 192.168.X.X and Subnet mask)

Client

Network computer or software that requests applications from a "host" or server.

Codec

General term for hardware and software combinations that compress and decompress video images or other file types.

Compression

Software algorithms that reduce the size of image files to quicken transmission time and reduce storage volume requirements. Often the computer used to view images from a camera must have specific compression software installed for functionality. Some compression programs in use include JPEG4, MPEG, and others. Compression software is not standardized . . . different vendors use different programs.

Crossover Cable

UTP jumper cable that is used for programming of network devices by direct connection to a laptop or desktop PC. Also used to connect Wi-Fi Access Points. W/Orange crosses to W/Green, Orange crosses to Green. What is "transmit" at one end goes to "receive" on the other.

DDNS—Dynamic Domain Name System

Web service that tracks changes in Dynamic IP addresses. Authorized users can sign onto the DDNS over the Internet and obtain an updated IP address for a location. A client application must be installed in the local PC at the address that will periodically contact the DDNS and provide the updated IP address.

DHCP—Dynamic Host Configuration Protocol

Network program that allows connected devices to automatically obtain a unique IP address, along with the network's common default gateway address and subnet mask. Commonly used by ISPs to provide WAN IP addresses to xDSL and cable modem adapters. Also used in Wi-Fi and Ethernet LANs. DHCP functionality is commonly included in LAN network routers as a programmable option, "on" or "off." If "on," the router provides DHCP "leases" of usable IP addresses to connected computers and devices.

Dial-Up

Accessing the Internet using a telephone line and modem. Connections are made via telephone calls into modem banks provided by an ISP.

DNS—Domain Name Server

Web server that provides translation of names (www.google.com) into numeric addresses (45.32.35.11). Typically provided as a part of an ISP's service package.

DSL—Digital Subscriber Line

Always-on adapter that provides two-way communications with the Internet via extra bandwidth in standard telephone lines. DSL is only available if the home or office is within 10–15,000 feet of a properly equipped telephone company central office.

DVR—Digital Video Recorder

Combined computer hardware and software that typically accepts analog camera inputs, digitizes the video signals, and stores them onto the DVR's hard drive(s). Features such as the number of cameras supported, alarm input/output, and recording options can vary, based on manufacturer and model. Some DVRs are network-capable, usually providing a female RJ-45 connector allowing wired Ethernet connection. Network DVRs will function as a "web server," with authorized users being able to access the DVR using standard web browser software such as Internet Explorer and Netscape.

Dynamic IP Addressing

See DHCP.

Enterprise

Network or data on the network that is associated with a client's business or primary function. Enterprise data may include VoIP telephony, data communications, email, and interoffice traffic. Also encompasses entire network architecture.

Ethernet

802.3 protocol standard for communication on wired networks. It is the most popularly used computer network communication standard. Ethernet-enabled devices can transmit and receive data at 10, 100, or 1,000 megabits per second.

Firewall

Hardware and/or software that protects nodes from unwanted intrusion by other computers on the network and/or Internet. Firewalls check the leading packets of information coming in, and either allow or disallow the traffic based on the authorization or authentication of the source. Firewalls have adjustable settings to allow "hosted applications" such as online gaming and connection of net cameras to be viewed over the Internet.

Gateway

Another term for a router, connects two networks together. In a system with a router, the "Default Gateway" is the IP address of the router.

Host

A computer on the network that provides programs to others. See "Server."

Hub

"Dumb" device that connects multiple cables together into a segment. Packets reaching a hub are rebroadcast to all connected ports. Bandwidth is shared between all connections. Because hubs retransmit data they can be used to increase the cabled distances between network devices.

IANA—Internet Assigned Number Authority

Organization responsible for assigning IP addresses on the Internet.

IEEE—Institute of Electronic & Electrical Engineers

Standards-setting organization for Ethernet (802.3) and Wi-Fi (802.11a, b, and g).

Industrial Ethernet

General term describing the connection of industrial equipment to a wired Ethernet network. Input/Output connections on the machinery, usually RS-232, 422, or 485, are connected to a module which provides a wired or fiber optic connection to an Ethernet network, web server functions, and IP addressability.

Internet

A network of networks, spanning the globe, which provides users with a wide variety of services. Internet access is usually provided by an ISP, which allows users to connect to the Internet via dialup or broadband.

IP (Internet Protocol) Address

Number, in a specific format, which uniquely identifies each device on an Ethernet and/or Wi-Fi network. Addresses are currently formatted in four octets, such as 192.168.101.212. In a valid IP address, each octet cannot exceed 255 in value. (See also Class of Network and Subnet Mask)

ISP—Internet Service Provider

Large network companies such as Yahoo, AOL, Earthlink, and others that provide users access to the Internet through dialup and broadband connections.

LAN—Local Area Network

A connected group of computers, printers, and other network devices that share a common network class IP address, for example 192.168.0.XXX. Generally used to describe connected computers, printers, etc. within a home, building, or campus of buildings.

LAN (or "Local") IP Address

Term used to differentiate the local IP address from the WAN (Internet) IP address. Example: the desktop computer has a LAN IP address of 192.168.1.101, while the Internet IP address for the network is 68.23.161.7. (See also WAN IP address.)

MAC (Media Access Control) Address

Equipment address programmed into network interface products at the factory. Used as an identifier of equipment on the network, comparable to a serial number.

Mbps—Megabits per Second

1,000,000 bits per second Unit of measurement for the throughput of network data.

Medium

This refers to the cable that carries computer signals. These used to be primarily coax, but now are usually four-pair Unshielded Twisted Pair-"UTP." Can also be fiber optics or RF signals when using Wi-Fi.

Modem

Compressed word from "Modulator—Demodulator." Modem sets convert and reconvert digital transmission into analog for transmission over standard phone lines.

NAT—Network Address Translation

Routers and proxy servers translate the external or WAN IP address into internal address(es) inside the network so that "hosted applications" can be reached from the Internet, and vice versa. Routers and firewalls must be specifically programmed to allow connections to IP devices inside the network.

NVMS—Network Video Management Software

Software sets that generally provide a viewing, control, and recording platform for a quantity of network-enabled video cameras or video servers.

NVR—Network Video Recorder

Industry term for a hardware and/or software device that provides recording and playback of network video images and files.

NIC—Network Interface Card

Adapter, either built-in or a separate card, which provides an electrical connection (usually an RJ-45 female outlet) for connection to an Ethernet network. Each NIC has a separate MAC address. Every device on an Ethernet network must have an NIC.

Node

Addressed device on the network. PCs, printers, net cameras, and DVRs are examples of nodes.

Packet

Standardized format for the transmission of data on Ethernet and Wi-Fi networks. Files are transmitted by being broken into packets of specific bit amounts, including the receiver's and sending computer's IP addresses, packet number, and error correction. Packets can travel by various routes to the receiving computer, which reassembles the packets into their original order.

Ports

Software channels used by TCP/IP networks for different specific services. Similar to the channels on a television set. Ports must be "open" (not blocked) by the router and/or firewall to allow incoming traffic, such as requests to view a network camera. Ports must be "forwarded" by the router to the network device's IP address.

Port Forwarding

Using programmable NAT within a router to forward requests and communications to a network device on the LAN.

Private IP Address Ranges

The following address ranges have been designated by the IANA (Internet Assigned Numbers Authority) as "private" IP address ranges to be used for non-Internet, i.e. local, network address ranges.

Class	Address Range	Default Subnet Mask
A	10.0.0.0–10.255.255.255	255.0.0.0
B	172.16.0.0–172.31.255.255	255.255.0.0
C	192.168.0.0–192.168.255.255	255.255.255.0

Network administrators will select from these ranges to address their local networks.

Protocol

Standardized sets of rules for communications between devices. Ethernet and Wi-Fi have specific rules for connection, packet size, acknowledgements, error correction, and other issues to enable communications between devices on networks.

Proxy Server

Gateway device or software that performs three functions. 1. Network address translation (NAT) and masking of the IP addresses of local computers. 2. Optionally caching of commonly requested web pages for quick presentation to requesting computers. 3. Logging of incoming and outgoing data packets and the computers which generated or received them.

Public IP Address

The Internet IP address of a network.

Router

Also called a "gateway router." Programmable device, connects different networks together, such as a LAN and a WAN (the Internet). Routers have two IP addresses, one for the LAN and one for the WAN. Routers shield outside networks from internal addresses and provide Network Address Translation (NAT) for packets allowed into the local network from the Internet. Routers can provide some levels of "firewall" protection.

Segment

Common cable or hub connection for one or more wired network devices.

Server

Computer that holds application programs and data and serves them to client computers when requested. Servers give applications and data, while clients receive. A network-enabled camera is a server to the client, which is the computer that is requesting images.

SMTP—Simple Mail Transfer Protocol

Common protocol used to transmit email messages.

SSID—Service Set Identifier

Unique network name programmed into all Wi-Fi devices on the same network.

Static IP Address

A non-changing IP address, either on a LAN or WAN. If clients want to remotely access hosted applications, such as network DVRs or cameras, without having to check for or track a changed IP address, they should have their ISP provide a Static IP address.

Subnet Masking

A specific set of bit flags that indicate the type of network to which an IP-addressed device is directly connected. The default subnet mask for LAN devices is 255.255.255.0, indicating with the last octet ".0" that this device is connected to a Class C (local) network. Various subnet masks can be used by system administrators to divide a number of connected devices into "sub networks" which can only communicate within their own sub network address range. For example, a device with an IP address of 192.168.2.80 with a subnet of 255.255.255.192 will only communicate with other addresses on the network within the range of 192.168.2.64 to 192.168.2.127.

Switch

Connects nodes in a segment. Switches "learn" what devices are connected to them. When packets reach a switch, they are retransmitted only to the specific addressee. Provides higher bandwidth than a hub.

TCP/IP

Transmission Control Protocol/Internet Protocol. All of the Internet and most Ethernet networks use this protocol to control communications between networked devices.

URL—Uniform Resource Locator

"Name" address of a web page or resource on the Internet. A typical URL is http://www.SecurityNetworkingInstitute.com.

UTP—Unshielded Twisted Pair

Common term for copper network cabling, consisting of four pairs of individually insulated 24-gauge twisted solid copper wire, with an overall outer jacket. The twisting helps reduce the effects of EMI and RFI. Also commonly called "Cat 5," "Cat 5e," or "Cat 6."

VoIP—Voice over Internet Protocol

Technology that converts analog telephone communications into IP-addressed data packets and transmits them over a LAN, WAN, or the Internet.

WAN—Wide Area Network

A network comprising interconnected WAN and LAN networks. WAN is a general term, usually describing networks connected to other networks over long distances. The Internet is a WAN.

WAN IP Address

Internet IP address. Can be a "static" address or a "dynamic" address supplied by a DHCP server. This is the address used to remotely connect to network cameras and devices over the Internet. Also the IP address on the ISP side of a router. (See also LAN IP address.)

WAP—Wireless Access Point

Also called an "access point." Wi-Fi equivalent of a hub. Provides wireless coverage in a specific area. Must be connected to a hub or switch with UTP cable. Programmable for Wi-Fi features and capabilities. Adds Wi-Fi capability to any Ethernet network.

Web Server

Network device that can be accessed with a computer using standard web browser software such as Internet Explorer or Netscape.

WEP—Wired Equivalent Privacy

Data encryption standard for Wi-Fi. Each device on a WEP-enabled Wi-Fi network must have the WEP encryption key programmed into it to enable the device to be active on that network.

Wi-Fi

802.11 protocol standard for communication on wireless networks. No site license is required for installation. 802.11b provides a maximum of 11 Mbps, while 802.11g and 802.11a each provide a maximum of 54 MBPS. 802.11b and .g are currently the most popular.

Wi-Fi Router

Combined UTP/Wi-Fi network device, typically provides both wired and Wi-Fi capabilities. Commonly used in home and small business networks.

WLAN—Wireless LAN

A Wi-Fi network, or the Wi-Fi portion of a combined wired/wireless network.

APPENDIX B

Useful Commands for Troubleshooting Networks and Devices

By accessing the "command line" screen in a Windows-equipped PC, the technician can access and troubleshoot a variety of network issues.

To Reach the Command Line (Windows XP)

Click START, click RUN, and type COMMAND, press ENTER . . . "Command" window will open.

To Obtain the IP Addressing Information for a Computer

IPCONFIG/ALL (press ENTER key)—That particular PC's local IP address, subnet mask, default gateway, MAC address, and other information for all enabled network interfaces will be presented on the screen. Write down or print this information for later use or to restore the computer to its original settings.

To Determine whether the Computer's NIC Card is Functional

PING 127.O.O.1 (press ENTER key) This command performs a "loopback" test, verifying that the NIC in that particular PC is functional.

To Determine whether a Device is Connected to the Network

PING 192.168.1.123 (the LAN IP of the device in question) (press ENTER key)— PING sends out four data packets and asks for a response from the receiving network device. If PINGing results in no response ("Timed Out"), the device is not communicating with the computer due to a power outage, disconnection of a cable, or IP addressing errors.

To Trace the Communications Route to a Device

TRACERT 192.168.1.123 (the IP address of the device or service in question) (press ENTER key). This command will provide routing information, showing how data packets are being sent through the network.

Note

The PING and TRACERT commands can use either IP addresses (numeric) or URLs (names) as addresses. To see how an Internet-connected computer reaches a popular web site, you can type TRACERT(space)www.google.com. The first line of information will display the Internet IP (numeric) address of the website.

APPENDIX C

Internet Connection Information Websites

Note

The content or availability of listed Internet sites may change without notice. The author is not responsible for content or availability of these sites.

To Determine the Internet IP Address of a Device or Network

http://www.whatismyip.com—The WAN IP address of the network will appear on the computer.

To Set Up DDNS Services to Track a Dynamic IP Address

http://www.dyndns.org—This is a free service, but you must establish an account with a username and password.

Note

You must download to the client's PC a small application, called an "update client," such as "DynDns Updater." This software is available from links on the dyndns.org page. Once downloaded, the update client must be programmed with the DDNS service URL, username & password, and other information. Once completed, the update client will periodically check the Internet IP address and update the DDNS server of any changes.

To Test a Computer and/or Network for Security Problems

http://www.grc.com—This web site provides the "Shields Up!" free security test for Internet-connected computers.

APPENDIX D

Web Pages of Interest

www.Veo.com—Veo camera products. Information and purchasing.

www.panasonic.com/cctv—Panasonic CCTV web site. Tutorials on products.

www.howstuffworks.com—Great free website with detailed yet simple explanations of computer devices and networking.

www.axis.com—Axis is a large manufacturer of network cameras. Site includes product information and live network camera views of New York City traffic, and other locations around the world.

www.bluenetvideo.com—Video servers, demonstrations.

www.mobi-inc.com—ReCam software, network cameras.

APPENDIX E

Reference Books

The Complete Reference: Internet, 2nd Edition. McGraw-Hill, 2002.
Linksys Networks, The Official Guide, 2nd Edition. McGraw-Hill, 2003.
Jeff Duntemann's Drive-By Wi-Fi Guide. Paraglyph Press, 2003.
The Essential Guide to Networking. Prentice Hall, 2001.

INDEX

10/100 Mbps Ethernet, 13–14
568A connectors, 30–31
568B connectors, 30–31, 251
802.11 standards, 46, 251
802.11a standard, 46
802.11b standard, 46, 51
802.11g standard, 46

A
access control communications,
    107–108
access points, 47, 260
active Ethernet splitters, 141–142
adapters, 24
    accessing, 157–158
    defined, 252
    DMZ mode, 161–162
    firewalls, 159–162
    functions of, 157
    hosted applications, 159–161
    password access, 158
        settings, 162–163
ad hoc mode, 48–49, 195–201, 252
alarm transmitters, 229–234
    benefits of, 229
    costs of, 231
    digital dialer backup, 232
    installing, 231
    IP addressing, 231
    programming, 231–232

receiver compatibility, 230–231
    remote downloading, 232
    security issues in, 233
    selection of, 230
    signal encryption, 233
    signals, 231
    testing, 232
    transmission options, 233
analog cameras, 111–113
analog communications, 7–9
asymmetrical DSL (ADSL), 65
asynchronous data communications,
    103–104
attenuation, 10

B
backbone cabling, 32, 94–95
Back Orifice virus, 147–148
backups, 155
bandwidth, 123–128
    See also Internet
    allocation of, 91
    availability of, 123–124
    calculation of, 127–128
    controls, 90, 123
        compression percentages,
        126–127
        frames per second (fps),
        124–125
        image scaling, 125–126

defined, 252
    usage of, 127
baseband frequency, 9
bi-directional predictive frames
    (B-frames), 120
binary codes, 9–10
bit, 252
bit rate, 130
bit sampling, 117
Blowfish encryption program, 233
BlueNetVideo server, 207–210
Bootstrap Protocol, 181
bridges, 24, 252
brightness, 130
broadband, 64
    See also Internet
    defined, 252
    and electronic security, 66–67
    satellite, 67
    T1 leased lines, 67
brute force attacks, 146–147
bus topology, 28

C
cable modems, 66, 252
cables, 27–35
    categories, 34
    crossover, 34
    installation and performance, 35
    maximum distance, 33

cables (*continued*)
  patch cords, 34
  standards, 28–29
cabling, 93–102
  backbone, 94–95
  Ethernet, 12
  horizontal, 95
  horizontal cross-connect, 95
  standards, 4, 93
  structured, 93–94
  testers, 243
caching, 26
cameras, 109–113
  ad hoc mode, 195
  analog, 111–113
  Ethernet, 177–186
  IP addresses, 195
  movement of, 130–131
  network, 109–114
  Wi-Fi, 187–192
  in wireless laptop
    surveillance, 194–195
captions, 217
Category 5 cables, 34
Category 5e cables, 34
Category 6 cables, 34
CCTV cameras, 40–41, 107–108
chromatic blocks, 118
Class A networks, 18
Class B networks, 18
Class C networks, 18–19, 251
client, 253
clients, 26
codecs, 110, 253
collisions, 13
color television tubes, 115–116
communication rooms,
    security of, 102
communications, 7–10
communications networks. *See*
    networks
communications testing, 245–246
COM port, 104
composite video, 116
compression, 253
computer viruses, 147
connectors, 29–32
contrast, 130

crimping tools, 243
cropping, 215
crossover cables, 34, 253

D
DC pickers, 141–142
decimation, 118
default gateway, 73
denial of service (DoS) attacks,
    148–149
destination addresses, 16
dial-up Internet, 64, 254
digital communications, 9–10
digital communicators, 239–240
digital dialers, 240
digital signals, 116–117
digital subscriber line (DSL), 64–65
  See also Internet
  adapters, 157–163
  defined, 254
  frames per second (fps), 126
  image scaling, 126
  IP addresses, 80–81
digital subscriber loop access multi-
    plexer (DSLAM), 65
digital video recorders (DVR),
    221–227
  See also video control and
    recording
  benefits of, 226–227
  connections, 221
  controls, 224
  defined, 254
  desktop, 225–226
  disk storage in, 224
  evidentiary options in, 224
  functions, 221
  hacker security, 225
  network, 225
  recording options, 221–223
  video motion detection in,
    223–224
  viewing options in, 224
digitizing, 116
direct laptop testing, 246
DMZ mode, 161–162, 174–175
Domain Name Servers (DNS), 58
  defined, 254

for Wi-Fi cameras, 191
for Wi-Fi routers, 170
downstream data, 126
dual ports, 249
duplex communications, 2
Dynamic Domain Name Servers
    (DDNS), 58–59
  defined, 253
  services, 263
  settings, 191
Dynamic Host Configuration
    Protocol (DHCP), 56–58
  defined, 253
  IP addressing, 79
  settings, 171–172, 182

E
EIA/TIA 568 standards, 28
  maximum distances of cables, 32
  for structured cabling, 93
  types of cable, 29
  types of connectors, 29–32
electrical-to-fiber converters, 40
electron guns, 116
email messaging, 131–132, 192
employee screening, 155
encryption, 47, 216
Endpoint power switch, 139–140
enterprise network, 89–90, 254
Ethernet, 11–14
  cabling, 12
  collisions, 13
  computer hardware, 12
  converters, 105
  copper connections, 33–34
  defined, 254
  device addressing, 16
  devices, 21–26
  and electronic security, 14
  firmware, 12
  Gigabit, 14
  individual addressing in, 12–13
  IP addressing, 17–20
  MAC addressing, 17
  media converters, 41–42
  in network communications, 5
  nodes, 13
  overview, 11–12

packets, 15–16
segments, 13
serial communications, 105
serial servers, 105
standards, 13–14
tools, 243
transmissions, 13, 15
and Wi-Fi, 46
Ethernet cameras, 177–186
connecting to, 177–180
default IP address, 177–178
DHCP settings, 182
domain name servers (DNS), 182
FTP settings, 184–185
host name, 181
MAC search, 178–180
network addressing, 180–181
NTP time settings, 183–185
port address settings, 181
protocols, 181

F
Fast Ethernet, 33–34
fiber optics, 37–43
See also cables
analog transmission on, 40–41
in backbone cabling, 94–95
connections, 39
description of, 37–38
media converters, 100–101
in networking, 37
powering devices, 40–43
testing, 39–40
types of, 37–38
file encryption, 216
file transfer protocol (FTP),
132–133, 184–185
firewalls, 24–25
for adapters, 159–162
defined, 254–255
and hosted applications, 159–162
settings, 174
troubleshooting, 249
firmware, 12
flash drives, 153
frame check sequence, 16
frame rate, 130
frames, 15

frames per second (fps), 115,
124–126

G
gateway routers, 22–24, 83–86, 255
Gigabit Ethernet, 14
group of pictures (GOP), 120

H
hackers, 144–146
half-duplex communications, 2
head-end communications, 106
horizontal cabling, 32, 95
horizontal cross-connection, 95
horizontal decimation, 118
hosted applications, 159–161
host name, 73, 170
hosts, 55, 255
hubs, 22, 255
hue, 118

I
image scaling, 125–126, 215
in-camera storage, 131
industrial Ethernet, 255
information technology (IT)
management, 87–91
bandwidth allocation by, 91
and network security, 88–89
responsibilities of, 87–88
security concerns of, 88
working with, 89
inframes (I-frames), 120
interference, 10
Internet, 63
broadband, 64
cable modem, 66
defined, 255
development of, 6
dial-up, 64, 254
digital subscriber line (DSL),
64–65
in electronic security, 66–67
frames per second (fps), 126
and image scaling, 126
and network classes, 18–19
and network security, 154
satellite, 67
T1 leased lines, 68

virtual private network (VPN),
68–69
Internet Assigned Number
Authority (IANA), 255
Internet Protocol (IP) addresses,
55–61
alarm transmitters, 232
and communications, 19–20
computer settings, 84
defined, 255
determination of, 263
digital subscriber line (DSL),
80–81
domain name servers (DNS), 58
dynamic, 56–58, 79
dynamic domain name servers
(DDNS), 58–59
Ethernet, 17–18
host, 55
IPCONFIG command, 74
LAN/WAN networks, 83–86
network address translation
(NAT), 59–61
and network classes, 18–19
network-connected devices,
71–74
overview, 71
PING command, 74
ports, 59–61
private, 257
reuse of, 19
server, 55
static, 55, 79
voice over IP (VoIP), 81–82
Wi-Fi gateway router, 82–83
in Windows, 74–77
Internet service providers (ISPs),
64, 256
IP alarm transmitters, 229–234
benefits of, 229
costs of, 231
digital dialer backup, 232
installing, 231
IP addressing, 231
programming, 231–232
receiver compatibility, 230–231
remote downloading, 232
security issues in, 233

IP alarm transmitters (*continued*)
  selection of, 230
  signal encryption, 233
  signals, 231
  testing, 232
  transmission options, 233
IPCONFIG command, 72–74, 261
IP video servers, 207–210
  DDNS settings, 208
  image options, 209
  image settings, 208–209
  IP addresses, 208
  programming access in, 207
  serial port settings, 210

J
JPEG compression, 119
jumpers, 34

K
keystroke logging, 154

L
LAN communication testing, 247
laptops, 243–244
latency, 120
line interactive UPS, 137
local area networks (LANs), 11–14
  cabling, 12
  collisions, 13
  computer hardware, 12
  defined, 256
  firmware, 12
  IP addressing, 12–13, 85–86, 256
  nodes, 13
  segments, 13
  standards, 13–14
  transmissions, 13
logical networks, 245
logic bombs, 148
lossy compression, 117
luminance blocks, 118

M
mainframe networks, 3–4
Mavix Media Racer network servers, 107
media access control (MAC), 17, 170, 256

media converters, 40–43, 94–95
medium, 21, 256
Midspan power switch, 140
modems, 26, 64, 256
motion detection, 217, 223–224
moveable cameras, 52
MPEG2 compression, 120
MPEG compression, 119–120
multimode fiber, 39, 101

N
nailed down configurations, 106
National Television Standards Committee (NTSC), 116
network address translation (NAT), 59–61
  defined, 256
  port forwarding, 172–173
  in proxy servers, 26
network cameras, 109–114
  10/100 compatibility, 101–102
  benefits of, 109–110
  cabling, 95–96
  connection to network, 97
  equipment, 151–152
  fiber hookup, 100
  functions of, 110
  limitations of, 110–111
  media converters, 100–101
  movement of, 130–131
  parallel network, 98–99
  power sources, 96–97
  pre-installation testing, 101
  recording options, 113
  viewing options, 101
network-connected devices, 71–74
network interface card (NIC), 12, 21, 257
Network Properties window, 75–77
networks, 1–6
  cabling, 4
  classes, 18–19
  description of, 1
  development of, 1
  and electronic security, 6
  Internet. *See* Internet
  mainframe, 3–4
  personal computers, 4–5

protocols, 5
telephone, 2
network security, 144–156
  backups, 155
  communication lines, 152
  functions of, 144
  hackers, 144–146
  holes in, 150–151
  importance of, 143
  law enforcement in, 143–144
  planning, 151
  protection from inside attacks, 153–155
  protection from outside attacks, 152–153
  protection of video data, 155
  redundant systems in, 155
  regular review of, 156
  testing, 263
  threats, 146–150
    brute force, 146–147
    computer viruses, 147
    denial of service, 148–149
    logic bombs, 148
    spoofing, 149–150
    Trojan horses, 147–148
    worms, 148
  tools in, 243–244
network servers, 111–113
Network Time Protocol (NTP), 183–185
network video management program (NVMP), 133–134, 256
network video recorder (NVR), 133–134, 257
network video system, 109–113
  analog cameras, 111–113
  cameras, 109–111
  monitoring options, 113
  network servers, 111–113
nodes, 13, 21, 257
node type, 73

O
on-line UPS, 137
optical loss testing, 40
optical time domain reflectometer (OTDR), 40

P

packets, 15–16, 257
pan command, 130–131
parallel network, 89–90, 97–99
password cracking, 146–147
password protection, 154
patch cords, 34
personal computers, 4–5
physical address, 73
picture elements, 116
PING command, 261
pin out configuration, 29
polling, 66–67
portable monitoring
        stations, 52–53
port forwarding, 172–173, 257
ports, 59–61, 257
powering devices, 40–43
    network, 139–142
    overview, 135–136
    remote control and reset, 138
    uninterruptiblepower supply
        (UPS), 136–138
    Power over Ethernet
        (PoE), 139–142
        See also Ethernet; Voice over
            Internet Protocol (VoIP)
        advantages of, 141
        description of, 139
        growth of, 141
        intelligent discovery in, 141
        power suppliers in, 139–140
        and security systems, 142
        tapping modules for, 141–142
predicted frames (P-frames), 120
protocols, 5, 258
proxy servers, 26, 249–250, 258
punch tools, 243

R

ReCam software, 211–219
    camera programming in, 212
    captions selections in, 217
    file encryption in, 216
    functions of, 211
    live camera viewing in, 218–219
    motion detection in, 217
    recording options, 215–216

scheduling in, 218
video and audio compression in,
        213–214
receivers, 7–8
redundant systems, 155
residential CCTV, 52
resolution, 130
resolution lines, 116
RJ-45 connectors, 29
routers, 22–24
    accessing, 166
    defined, 258
    description of, 166
    DHCP settings, 171–172
    disabling SSID transmission, 169
    DMZ mode, 174–175
    DNS addresses, 170
    filtering options, 173–174
    firewall rules, 174
    functions of, 16
    host name, 170
    LAN settings, 171
    MAC address, 170
    MAC cloning, 170
    port forwarding, 172–173
    SSID name, 167
    static DHCP, 172
    WAN settings, 169–170
    Wi-Fi, 48, 82–86, 165–176
    wireless settings, 166–167
    working with, 175
RS-422 protocol, 103–104
RS-485 protocol, 103–104

S

satellite Internet, 67
saturation, 118, 130
scaling, 118–119, 125–126
scheduling, 218
segments, 13, 21, 258
sequence code, 16
serial communications, 103–108
    Ethernet converters, 105
    head-end, 106
    RS-422 protocol, 103–104
    RS-485 protocol, 103–104
    serial servers, 105
serial servers, 105

serial tunneling, 106
servers, 25–26, 55, 258
service set identifier (SSID), 167
singlemode fiber, 39, 101
site license, 45
spatial redundancy, 119
spoofing, 149–150, 154–155
standby backup offline
        UPS, 136–137
star configuration, 28–29
static IP, 55, 79, 258–259
storage of images, 131
structured cabling, 4, 93–94
subnet mask, 20, 73, 259
surveillance systems,
        wireless, 193–205
    camera programming in,
        194–195
    description of, 193–205
    disk space requirements in,
        202–203
    equipment, 194
    interference in, 204
    laptop programming in,
        195–201
    onsite installation, 203–204
    recording software in, 201–202
    techniques in, 204–205
switches, 16, 22, 48, 259
symmetrical DSL (SDSL), 65
synchronous data
        communications, 104

T

T1 lines, 68
TCP/IP ports, 60
telecommunication closet, 95
telephone lines, 2, 64
telephone networks, 2
television tubes, 115–116
testing, 245–248
    communications, 245–246
    direct laptop, 246
    LAN communication, 247
    WAN/Internet communications,
        247–248
thumb drives, 153
tilt command, 130–131

TL-250 IP transmitter, 230
token ring, 5
TRACERT command, 80, 262
Transmission Control
 Protocol/Internet Protocol
 (TCP/IP), 259
transmitters, 7–8
Trojan horses, 147–148
troubleshooting, 245–250
 logical networks, 245
 video images, 248–250

U
uniform resource locator
 (URL), 259
uninterruptible power supply
 (UPS), 136–138
 battery backup, 137
 capacity requirements, 138
 functions of, 136
 intelligent, 138
 line interactive, 137
 on-line, 137
 standby backup offline, 136–137
 use of, 43
unshielded twisted pair
 (UTP), 29, 259
update client, 263
upstream data, 126
UTP jumpers, 243

V
vampire taps, 27
vertical decimation, 118
video compression, 115–121
 analog to digital, 116–117
 in basic television, 115–116
 composite video, 116
 compression percentages,
 126–127
 decimation, 118
 image scaling and cropping, 215
 image stream, 115
 JPEG compression, 119
 lossless, 117
 lossy, 117
 MPEG compression, 119–120
 in networking, 121
 resolution lines, 116

scaling, 118–119
 spatial redundancy, 119
 temporal redundancy, 119
video control and recording,
 129–134
 camera movement in, 130–131
 compression compatibility,
 133–134
 email messaging in, 131–132
 file transfer protocol (FTP),
 132–133
 image control, 129–130
 image storage, 131
 image transfer, 133–134
 network, 133–134
 and network security, 155
 sensory overload in, 133–134
 software program, 133
 Web browser in, 133
video management software,
 211–219
 camera programming in, 212
 captions selections in, 217
 file encryption in, 216
 functions of, 211
 live camera viewing in, 218–219
 motion detection in, 217
 recording options, 215–216
 scheduling in, 218
 video and audio compression in,
 213–214
video motion detection (VMD),
 223–224
video servers, 207–210
 DDNS settings, 208
 image options, 209
 image settings, 208–209
 IP addresses, 208
 programming access in, 207
 serial port settings, 210
video spoofing, 149–150
viewing software, 249
Virtual Com Port software, 106
virtual private network (VPN),
 68–69
viruses, 147
Voice over Internet Protocol (VoIP),
 235–242

and alarm equipment, 240–242
 alarm transmission problems
 with, 239
 benefits of, 237
 connecting digital communicators
 to, 239–240
 dealers, 240–241
 defined, 259
 description of, 235–236
 installation problems with, 237
 IP addresses, 81–82
 power sources, 138–139
 service quality, 237

W
waveforms, 7
web browser, 133
web servers, 260
wide area networks (WANs), 85–86
 defined, 259
 IP addressing, 85–86, 260
 testing, 247–248
Wi-Fi networks, 52–53
 access points, 47
 ad hoc mode, 48–49
 bandwidth, 51–52
 bridges, 52–53
 coverage, 49
 data security in, 47
 defined, 260
 for electronic security, 52–53
 Ethernet emulation in, 46
 IP addressing, 82–83
 laptops, 48
 product compatibility in, 47
 routers, 48, 83–86, 165–176,
 260
 security of, 50–51
 shared frequencies, 46
 standards, 46
 switches, 48
 tools, 244
Wi-Fi wireless cameras, 187–192
 connecting to, 187
 DDNS settings, 191
 description of, 187
 DNS settings, 191
 email settings, 192

image settings, 191–192
IP address settings, 190
port settings, 190–191
Wi-Fi settings, 188–189
Wired Equivalent Privacy (WEP),
47, 167–168, 260
Wireless Access Point (WAP), 260
Wireless Ethernet Compatibility
Standard (WECA), 47

wireless LANs, 45–53, 260
wireless laptop surveillance,
193–205
camera programming in,
194–195
description of, 193–205
disk space requirements in,
202–203
equipment, 194

interference in, 204
laptop programming in,
195–201
onsite installation, 203–204
recording software in, 201–202
techniques in, 204–205

Z
zoom command, 130–131